Lóegaire Humphrey (Ed.)

Dobra Bridge (A6)

Lóegaire Humphrey (Ed.)

Dobra Bridge (A6)

A6 (Croatia), Gorski kotar, Dobra (river), Autocesta
Rijeka – Zagreb

Claud Press

Imprint

Permission is granted to copy, distribute and/or modify this document under the terms of the GNU Free Documentation License, Version 1.2 or any later version published by the Free Software Foundation; with no Invariant Sections, with the Front-Cover Texts, and with the Back- Cover Texts. A copy of the license is included in the section entitled "GNU Free Documentation License".

All parts of this book are extracted from Wikipedia, the free encyclopedia (www.wikipedia.org).

You can get detailed informations about the authors of this collection of articles at the end of this book. The editors (Ed.) of this book are no authors. They have not modified or extended the original texts.

Pictures published in this book can be under different licences than the GNU Free Documentation License. You can get detailed informations about the authors and licences of pictures at the end of this book.

The content of this book was generated collaboratively by volunteers. Please be advised that nothing found here has necessarily been reviewed by people with the expertise required to provide you with complete, accurate or reliable information. Some information in this book maybe misleading or wrong. The Publisher does not guarantee the validity of the information found here. If you need specific advice (f.e. in fields of medical, legal, financial, or risk management questions) please contact a professional who is licensed or knowledgeable in that area.

Any brand names and product names mentioned in this book are subject to trademark, brand or patent protection and are trademarks or registered trademarks of their respective holders. The use of brand names, product names, common names, trade names, product descriptions etc. even without a particular marking in this works is in no way to be construed to mean that such names may be regarded as unrestricted in respect of trademark and brand protection legislation and could thus be used by anyone.

Cover image: www.ingimage.com
Concerning the licence of the cover image please contact ingimage.

Publisher:
Claud Press is a trademark of
International Book Market Service Ltd., 17 Rue Meldrum, Beau Bassin, 1713-01 Mauritius
Email: info@bookmarketservice.com
Website: www.bookmarketservice.com

Published in 2012

Printed in: U.S.A., U.K., Germany. This book was not produced in Mauritius.

ISBN: 978-613-8-43989-9

Contents

Dobra_Bridge_(A6)

<table>
<tr><td colspan="2" align="center">Dobra Bridge</td></tr>
<tr><td colspan="2">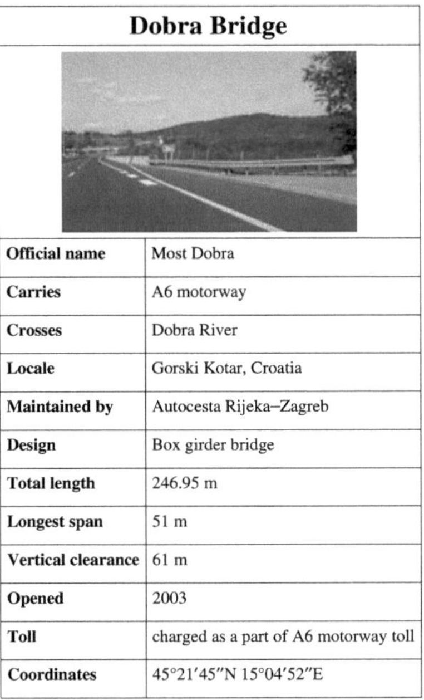</td></tr>
<tr><td>Official name</td><td>Most Dobra</td></tr>
<tr><td>Carries</td><td>A6 motorway</td></tr>
<tr><td>Crosses</td><td>Dobra River</td></tr>
<tr><td>Locale</td><td>Gorski Kotar, Croatia</td></tr>
<tr><td>Maintained by</td><td>Autocesta Rijeka–Zagreb</td></tr>
<tr><td>Design</td><td>Box girder bridge</td></tr>
<tr><td>Total length</td><td>246.95 m</td></tr>
<tr><td>Longest span</td><td>51 m</td></tr>
<tr><td>Vertical clearance</td><td>61 m</td></tr>
<tr><td>Opened</td><td>2003</td></tr>
<tr><td>Toll</td><td>charged as a part of A6 motorway toll</td></tr>
<tr><td>Coordinates</td><td>45°21′45″N 15°04′52″E</td></tr>
</table>

Dobra Bridge is located between the Vrbovsko and Ravna Gora interchanges of the A6 motorway in Gorski Kotar, Croatia, spanning Dobra River. It is 246.95 metres (810.2 ft) long and carries the motorway across five spans. The bridge consists of two parallel structures, both completed in 2003 by Primorje.[1] [2] The bridge is tolled within the A6 motorway ticket system and there are no separate toll plazas associated with use of the bridge. The bridge is located in a water protection area.[3]

Traffic volume

Traffic is regularly counted and reported by Autocesta Rijeka–Zagreb, operator of the viaduct and the A6 motorway where the structure is located, and published by Hrvatske ceste. Substantial variations between annual (AADT) and summer (ASDT) traffic volumes are attributed to the fact that the bridge carries substantial tourist traffic to the Adriatic resorts. The traffic count is performed using analysis of motorway toll ticket sales.[4]

Dobra Bridge traffic volume				
Road	Counting site	AADT	ASDT	Notes
A6	3006 Vrbovsko west	11,979	20,091	Between Vrbovsko and Ravna Gora interchanges.

See also

- List of bridges by length

References

[1] "Spojen desni most Dobra [Right hand side of Dobra Bridge structure joined]" (http://novine.novilist.hr/default.asp?WCI=Rubrike&
WCU=28592863285C2863285A28582858285B286328962897289E28632863285D285A285D2861285D2863286328632863A) (in Croatian).
Novi list. April 1, 2003. . Retrieved October 12, 2010.

[2] "Viadukti in mostovi [Viaducts and bridges]" (http://www.primorje.si/index.php?vie=cnt&str=49_slo) (in Slovenian). Primorje d.d.. .
Retrieved October 12, 2010.

[3] (PDF) *Croatian Motorways* (http://www.hac.hr/files/file/brosure/monografija/virtualMagazine.html). Hrvatske autoceste. 2007.
pp. 352–353. ISBN 9789537491. . Retrieved September 11, 2010.

[4] "Traffic counting on the roadways of Croatia in 2009 - digest" (http://www.hrvatske-ceste.hr/WEB - Legislativa/brojenje-prometa/
CroDig2009.pdf). *Hrvatske Ceste*. May 1, 2010. .

A6_(Croatia)

A6 motorway	
Autocesta A6	
Autocesta Rijeka–Zagreb	
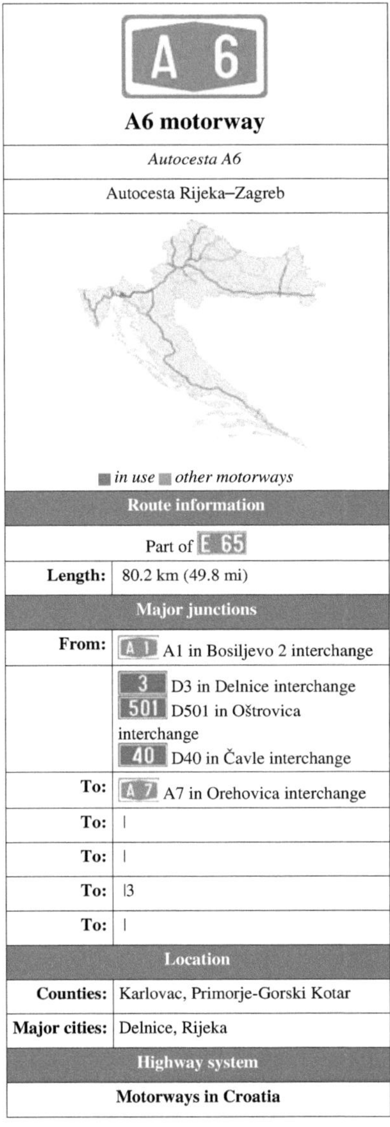	
■ *in use* ■ *other motorways*	
Route information	
Part of E 65	
Length:	80.2 km (49.8 mi)
Major junctions	
From:	A1 A1 in Bosiljevo 2 interchange
	3 D3 in Delnice interchange 501 D501 in Oštrovica interchange 40 D40 in Čavle interchange
To:	A7 A7 in Orehovica interchange
To:	I
To:	I
To:	I3
To:	I
Location	
Counties:	Karlovac, Primorje-Gorski Kotar
Major cities:	Delnice, Rijeka
Highway system	
Motorways in Croatia	

The **A6 motorway** (Croatian: *Autocesta A6*) is a motorway in Croatia spanning 80.2 kilometres (49.8 mi).[1] It connects the nation's capital, Zagreb, via the A1, to the seaport of Rijeka.[2] The motorway forms a major north–south transportation corridor in Croatia and is a part of European route E65 Nagykanizsa–Zagreb–Rijeka–Zadar–Split–Dubrovnik–Podgorica. The A6 motorway route also follows Pan-European corridor Vb.[3]

The A6 motorway runs near a number of Croatian cities, provides access to Risnjak National Park and indirectly to numerous resorts, notably in the Istria and Kvarner Gulf regions. The motorway route was completed in 2008. The motorway is nationally significant because of its positive economic impact on the cities and towns it connects, and because of its contribution to tourism in Croatia.[4] The importance of the motorway as a transit route will be further increased upon completion of a proposed expansion of the Port of Rijeka and Rijeka transport node.[5] [6] [7]

Bosiljevo 2 interchange, the northern terminus of the A6 motorway

The motorway consists of two traffic lanes and an emergency lane in each driving direction separated by a central reservation. Sections of the motorway that have a gradient greater than 4% are divided into three lanes to prevent traffic problems caused by slower vehicles. These sections have no emergency lanes. Similarly, there are no emergency lanes in the tunnels. All intersections of the A6 motorway are grade separated. As the route traverses rugged mountains it requires numerous long bridges, viaducts, tunnels, and other structures. As of 2010 there are nine exits and three rest areas situated along the route.[8] The majority of the motorway is a ticket system toll road with pricing tied to vehicle classification. Each exit between Grobnik mainline toll plaza and Bosiljevo 2 interchage has a toll plaza. No toll is charged at Bosiljevo 2 where the traffic switches to the A1 motorway; traffic is tolled upon leaving the A1 motorway. Exits between the mainline toll plaza and Orehovica interchange have no toll plazas, as that part of the A6 route is not tolled.[9]

A motorway connecting Zagreb and Rijeka was originally designed in the early 1970s, and construction started north of Rijeka and south of Zagreb. The first section, between Rijeka and Kikovica, opened on September 9, 1972, and a Zagreb–Karlovac section followed on December 29, 1972. Those sections were the first modern motorways to be built in Croatia and Yugoslavia.[10] Due to political upheavals in Croatia and Yugoslavia, construction of the motorway was labeled a "nationalist project" and, along with the proposed Zagreb–Split motorway, was cancelled in 1971.[11] After the Croatian War of Independence, efforts to build the motorway were renewed and construction resumed in 1996. In 2004, a two-lane, single carriageway expressway was completed between the sections completed 25 years previously, and the second carriageway was built; the motorway was completed on October 22, 2008. Construction costs are estimated at 661.5 million euro.[12] Although Hrvatske autoceste normally designs, builds, and operates motorways in Croatia, the A6 motorway is operated and maintained by Autocesta Rijeka – Zagreb.[13] [14]

Route description

Orehovica interchange, the southern terminus of the A6 motorway

The A6 motorway is a significant north–south motorway in Croatia connecting the largest seaport of the country, Rijeka, to its hinterland and to the rest of the Croatian motorway network via the A1 motorway Bosiljevo 2 interchange.[15] The motorway follows a route through the Gorski Kotar region. Part of the road network of Croatia, the motorway is also part of European route E65 Nagykanizsa–Zagreb–Rijeka–Zadar–Split–Dubrovnik–Podgorica.[16] The motorway is of major importance to Croatia in terms of development of the economy; it is especially important for tourism and as a transit transport route. The road serves tourist resorts in Istria and the Kvarner Gulf islands. Because of the link formed between Zagreb and Rijeka, tourism-related traffic originating from the countries neighbouring Croatia to the north flows via this road to the Adriatic coast on the south. The road also serves tourists originating in the northern inland areas of Croatia. Even though the A6 route predominantly follows an east–west orientation, the motorway is locally regarded as a north–south communication. The ultimate importance of the motorway as a transit route shall be achieved upon completion of the proposed expansion of Port of Rijeka and the Rijeka transport node. The expansion is planned to encompass an enhancement of the cargo handling capacity of the Port of Rijeka; improved railroad links; and a new Rijeka bypass motorway linking the A6, via a new interchange, with the present routes of the A7 and A8 motorways. One of the aims of the project is to increase traffic along the A6 route.[5] [6] [7] [17] As of the June 1997 Pan-European Transport Conference in Helsinki, the motorway is a part of the Pan-European corridor Vb.[3] [18]

The motorway spans 80.2 kilometres (49.8 mi) between Bosiljevo 2 interchange and Rijeka–Orehovica interchange on the A7 motorway. The route serves Vrbovsko via the D42, Delnice via the D3, Crikvenica and Krk via the D501, and Bakar via the D40 state road. The route is complete and further development of the motorway includes only the construction of additional rest areas.[2] The A6 motorway consists of at least two traffic lanes and an emergency lane in each driving direction along its entire length, except in tunnels, where there are emergency bays instead. Sections of the A6 motorway steeper than 4% grade have three traffic lanes, and slow vehicles are restricted to driving in the rightmost lane. All of the interchanges are trumpet interchanges. There are a number of rest areas along the motorway providing various types of services ranging from simple parking spaces and restrooms to filling stations, restaurants, and hotels.[8] [19] As of October 2010, the motorway has nine interchanges providing access to numerous towns and cities and the Croatian state road network.[20] The motorway is operated by Autocesta Rijeka–Zagreb.[13]

An automatic traffic monitoring and guidance system is in place along the motorway. It consists of measuring, control, and signalling devices, located in zones where driving conditions may vary—at interchanges, near viaducts, bridges, tunnels, and in zones where fog and strong wind are known to occur. The system comprises variable traffic signs used to communicate changing driving conditions, possible restrictions, and other information to motorway users.[21]

The A6 motorway mainly runs through the mountainous Gorski Kotar region, requiring not only large bridges and viaducts and long tunnels along the route, but also special care must be paid to protection of the environment, as the route is located in karst

Variable traffic signs on the A6

terrain, with numerous water supply protection zones and significant natural heritage.[22] Risnjak National Park is located near the A6 route, and is accessed via the Delnice interchange. Due to the motorway access and its proximity to a number of seaside resorts, Risnjak is the most visited national park in Croatia.[23] Karst terrain is especially susceptible to water pollution, so the A6 motorway is equipped with a closed water drainage system designed to channel rainwater, meltwater, and any spillages to purpose-built processing facilities. Approximately 200 karst features—caves and other types of karst features—were observed and protected during construction of the motorway. An extraordinary example of this was a cavern 83 m (272 ft) long by 63 m (207 ft) wide and 45 m (148 ft) tall, found during execution of the 260 m (850 ft) long Vrata Tunnel.[24] The cavern was bridged by one of the tunnel tubes, which was sealed to protect the cavern and the water flowing through it.[22] [25] [26]

Toll

The A6 is a tolled motorway based on the vehicle classification in Croatia using a closed toll system integrated with the A1 motorway. The two roads connect at the Bosiljevo 2 interchange, forming a unified toll system. Since the A1 motorway is operated jointly by Autocesta Rijeka–Zagreb and Hrvatske autoceste, the toll collection system is operated jointly by the two operators.[27] As of October 2010, the toll charged along the A6 route between Bosiljevo 2 interchange (A1 Bosiljevo exit) and the Kikovica mainline toll plaza varies depending on the length of route travelled and ranges from 6.00 kuna (0.82 euros)

Oštrovica toll plaza

to 33.00 kuna (4.52 euros) for passenger cars and 25.00 kuna (3.42 euro) to 139.00 kuna (19.04 euro) for semi-trailer trucks.[9] The toll is payable in either Croatian kuna or euros and by major credit cards and debit cards. A number of prepaid toll collection systems are also used, including various types of smart cards issued by the motorway operator and *ENC*—an electronic toll collection (ETC) system which is shared by most motorways in Croatia and provides drivers with discounted toll rates for dedicated lanes at toll plazas.[27] [27]

The toll collected by Autocesta Rijeka–Zagreb for use of the A6 motorway is not reported separately. Autocesta Rijeka–Zagreb only reports it total toll revenue, including toll revenue collected on the A7 motorway (Rupa–Jurdani section) and the A1 motorway (Lučko–Bosiljevo 2 section) as well as on the Krk Bridge. In the first half of the 2010 their toll revenue was 188.2 million Croatian kuna (25.3 million euros).[28]

Notable structures

Bajer Bridge

As the A6 motorway route runs through mountainous terrain of Gorski Kotar, it comprises a substantial number of major structures—bridges, viaducts, tunnels, underpasses, flyovers, and culverts. Out of the total length of the Rijeka–Zagreb motorway of 146.5 kilometres (91.0 mi), 22.1 kilometres (13.7 mi) are situated within such structures. The northern part of the Rijeka–Zagreb motorway, designated as the A1 motorway, comprising 38.6 kilometres (24.0 mi) between Zagreb and Karlovac, contains only 572 metres (1877 ft) of such structures as the section is situated in a plain.[29] The 11.4 kilometres (7.1 mi) between Karlovac and Bosiljevo 2 interchanges, contains as much as 4036 metres (13241 ft) of the structures.[30] Thus the A6 motorway has 17.5 kilometres (10.9 mi), or 21% of the route, located within such structures. The Rijeka–Zagreb motorway has a total of 24 viaducts, 13 tunnels, 5 bridges, 45 underpasses, and 26 flyovers. All of the bridges, viaducts, and tunnels on the A6 motorway have at least two driving lanes in each direction.[31]

The longest tunnel on the A6 motorway route is the 2143-metre (7031 ft) Tuhobić Tunnel, located on the Oštrovica–Vrata section. The tunnel was initially opened as a single-tube tunnel in 1996.[32] [33] The second tunnel tube was excavated in August 2007 and opened to traffic in 2008.[31] [34] The European Tunnel Assessment Programme (EuroTAP), a tunnel safety assessment programme supported by the European Commission, coordinated by FIA and led by the German motoring club ADAC, tested Tuhobić Tunnel twice—once in 2004, when it achieved poor results, and again in 2009 after implementation of EuroTAP safety recommendations. The 2009 test ranked the tunnel as the second safest in Europe.[35] An unusual

Podvugleš Tunnel

feature associated with the A6 tunnels is the close proximity of the 1490-metre (4890 ft) Javorova Kosa and the 610-metre (2000 ft) Podvugleš tunnels—they are separated by less than 60 metres (200 ft) of road. In order to prevent abrupt changes in road conditions caused by the weather, the distance between the tunnels is covered by translucent roofing.[25] The tunnels are located on the Vrbovsko–Ravna Gora section.[36] Other significant tunnels on the A6 motorway are the 1130-metre (3710 ft) Veliki Gložac and Vrata tunnels. While the former, as with all the other tunnels mentioned, is significant due to its length, the latter is notable for the large cavern encountered during its excavation.[22] [25]

The most significant bridges and viaducts on the A6 motorway route are the 485-metre (1591 ft) Bajer Bridge spanning Lake Bajer near Fužine, on the Vrata–Oštrovica section,[33] and the Zečeve Drage and Severinske Drage viaducts. The two viaducts are 924 metres (3031 ft) and 725 metres (2379 ft) long respectively. The remaining viaducts on the motorway that are longer than 500 metres (1600 ft) are Hreljin and Golubinjak viaducts.[25]

History

Tuhobić Tunnel, the longest tunnel on the A6 route

Transport links between Rijeka and Zagreb have always been of substantial importance because of the transport requirements of the Port of Rijeka. This was first recognised by the Habsburg Empire in 1728, when the *Carolina* road was completed, and again in 1780 when the road was modernized. The original Rijeka–Zagreb road was replaced in 1811 by a new route, the *Louisiana* road, in order to avoid the steep sections of its predecessor. The new road remained the primary transport link to Rijeka until 1873, when the first railroad to the city was built. Further development of the port and industry in Rijeka and Zagreb required a more efficient road, which was built in 1954. That road was to remain the principal road transport link between the two cities for decades.[25]

Zagreb–Rijeka motorway, of which the A6 motorway is a part, was one of three routes defined in 1971 as priority transport routes of Yugoslavia that were to be developed as motorways.[11] The first section of the A6 motorway, between Orehovica and Kikovica, was 10.5 km (6.5 mi) long and opened on September 9, 1972. The section was also the first six-lane motorway built in Yugoslavia. The 39.3-kilometre (24.4 mi) long Zagreb–Karlovac section, now designated the A1 motorway, was completed on December 29, 1972.[37] Further construction was suspended for the following 25 years, as a political decision had been made by the Yugoslav leadership to withdraw funding for the construction. The funds were instead allocated to the construction of a motorway that would travel between Ljubljana, Zagreb, Belgrade, and Skopje, then known as the Brotherhood and Unity Highway. The Croatian section of the highway later became the A3 motorway. After the breakup of Yugoslavia, construction of the Rijeka–Zagreb motorway was still on hold due to the Croatian War of Independence, and no further construction took place until 1996.[11] The sole exception to the 25-year-long hiatus was the 7.25-kilometre (4.50 mi) long Kikovica–Oštrovica section, which was originally executed as an expressway and opened in 1982.[10] [38]

In 1996, construction of the A6 motorway resumed, and in 1997, a further 30 km (19 mi) of expressway between Oštrovica and Kupjak was completed. In December 1997, the government of the Republic of Croatia founded the Autocesta Rijeka–Zagreb company and tasked it with operating the completed sections of motorway and the construction of the remainder of the route.[39] [40] The new motorway operator resumed construction in three stages. During the first stage, 60.18 km (37.39 mi) of expressway between Kupjak and Karlovac were completed by the end of June 2004, comprising 60.18 kilometres (37.39 mi) of motorway and semi-motorway.[38] In the second stage, the expressway was upgraded to a full motorway by the end of October 2008. This stage required additional construction along 55.57 kilometres (34.53 mi) of the route.[31] [41] The upgraded motorway was officially opened on October 22, 2008, by Prime Minister Ivo Sanader at a ceremony held at the southern portal of Tuhobić Tunnel.[12] The opening ceremony coincided with opening of a new bridge over the river Mura on the border between Croatia and Hungary, connecting the A4 to the Hungarian M7 motorway. Thus the route spanning Budapest–Zagreb–Rijeka was completed as a modern motorway.[42] [43] Construction costs incurred are estimated at 661.5 million euros.[12] Even though Hrvatske autoceste normally develops motorways in Croatia, the A6 motorway is operated and maintained by Autocesta Rijeka–Zagreb.[13] [14]

Traffic volume

Traffic is regularly counted by means of a traffic census at toll stations and reported by Autocesta Rijeka–Zagreb, the operator of the motorway, and published by Hrvatske ceste. The reported traffic volume exhibits no significant variations as the motorway chainage increases, and as it passes by various major destinations and the interchanges that serve them, except at the Vrata interchange, where traffic to and from Krk Island, Crikvenica, and Novi Vinodolski flows. The greatest volume of traffic is registered between Delnice and Vrata interchanges—with a 12,600-vehicle annual average daily traffic (AADT), and a 21,150-vehicle average summer daily traffic (ASDT) figure. Sections south of Kikovica interchange likely carry substantial

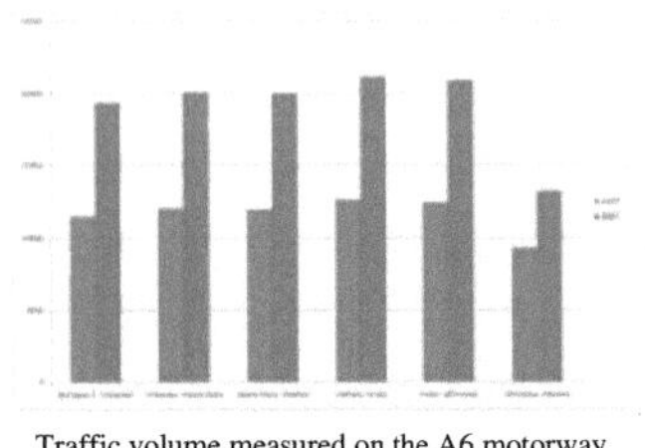

Traffic volume measured on the A6 motorway (2009)

traffic volume as they serve Rijeka commuter traffic as well as the volume registered between the Oštrovica and Kikovica interchanges. However, no traffic volume figures are published for those sections, since motorway traffic is counted by means of toll ticket sales analyses, and the sections south of Kikovica interchange are not tolled.[44]

Substantial variations observed between AADT and ASDT are normally attributed to the fact that the motorway carries significant tourist traffic to Istria and Kvarner Gulf. The seasonal increase in traffic volume ranges from 41% on the Oštrovica–Kikovica section to 69% as measured on the Bosiljevo 2–Vrbovsko section. The average summer-season traffic volume increase on the motorway is 65%.[4] [44]

A6 traffic volume details				
Road	**Counting site**	**AADT**	**ASDT**	**Notes**
A6	3022 Bosiljevo 2 west	11,448	19,401	Between Bosiljevo 2 and Vrbovsko interchanges
A6	3006 Vrbovsko west	11,979	20,091	Between Vrbovsko and Ravna Gora interchanges
A6	2906 Ravna Gora west	11,900	20,004	Between Ravna Gora and Delnice interchanges
A6	2910 Delnice west	12,600	21,150	Between Delnice and Vrata interchanges
A6	2915 Vrata west	12,413	20,891	Between Vrata and Oštrovica interchanges
A6	2933 Oštrovica west	9,324	13,168	Between Oštrovica and Kikovica interchanges

Rest areas

Approach to the Ravna Gora rest area

As of October 2010, there are four rest areas operating along the A6 motorway,[19] as a new rest area opened on October 9, 2010 next to the western portal of Tuhobić Tunnel on the Vrata–Oštrovica section of the route. Applicable legislation provides for four types of rest areas designated as types A through D: A-type rest areas comprise a full range of amenities including a filling station, a restaurant and a hotel or a motel; B-type rest areas have no lodging; C-type rest areas are very common and include a filling station and a café, but no restaurants or accommodations; and D-type rest areas offer parking spaces only, with possibly some picnic tables, benches, and restrooms.[45] Even though the rest areas found along the A6 motorway generally follow this ranking system, there are considerable variations, as some of them offer extra services. The most notable example is Lepenica rest area—even though it has no restaurant and therefore falls below B-type rest area standard, there is, for instance, an RV park available. The filling stations typically have small convenience stores and some of them offer LPG fuel. As of October 2010, all of the rest areas found along the A6 motorway comply with C-type rest area standards or above.[19]

The primary motorway operator, Autocesta Rijeka–Zagreb, leases the rest areas to various operators through public tenders. As of October 2010, there are three such rest area operators on the A1 motorway: INA, OMV and Tifon. The rest area operators are not permitted to sub-lease the fuel operations; the Tifon-operated rest area has a restaurant and a hotel operated by Marché, a Mövenpick Hotels & Resorts subsidiary, but they are also penalized if some facilities required by the lease contract are not operating.[46] All of the A6 motorway rest areas, except Ravna Gora, are accessible from one of the directions of the motorway traffic only. The rest areas normally operate operate 24 hours a day, seven days a week.[19]

List of A6 motorway rest areas				
County	km	Name[2]	Operators	Notes
Primorje-Gorski Kotar	28.8	Ravna Gora	Tifon Marché	Facilities found at Ravna Gora rest area comprise a filling station selling petrol, diesel fuel and LPG, a café, a restaurant, a hotel, free wireless network access, showers, and restrooms.[47] [48] The hotel and the restaurant are operated by Marché.[49]
	56.3	Lepenica	INA	Facilities found at Lepenica rest area comprise a filling station selling petrol and diesel fuel, a café, an RV park, and restrooms. Accessible to the southbound traffic only[50] [51]
	59.0	Tuhobić	OMV	As of October 2010, facilities found at Tuhobić rest area comprise a filling station selling petrol, diesel fuel and LPG. A restaurant is expected to open at the rest area in near future.[52] Accessible to northbound traffic only.[53]
	74.5	Cernik-Čavle	INA	Facilities found at Cernik-Čavle rest area comprise a filling station selling petrol and diesel fuel and restrooms. Accessible to the northbound traffic only[51] [54]
1.000 mi = 1.609 km; 1.000 km = 0.621 mi				
Concurrency terminus • Closed/former • Incomplete access • Unopened				

Exit list

County	km	Exit	Name[2]	Destination[16] [20]	Notes
Karlovac	0.0	1	Bosiljevo 2	A1 E65	Connection to the A1 motorway in Bosiljevo 2 interchange[2] The northern terminus of European route E65 concurrency The northern terminus of the motorway
Primorje-Gorski Kotar	4.4				Severinske Drage Viaduct
	9.0				Veliki Gložac Tunnel
	11.9				Zečeve Drage Viaduct
	15.8	2	Vrbovsko	D42	Connection to Vrbovsko via the D42 state road[20]
	16.1				Dobra Bridge
	17.5				Kamačnik Bridge
	26.5				Podvugleš Tunnel
	27.7				Javorova Kosa Tunnel
	28.8	P	Ravna Gora rest area		
	30.9	3	Ravna Gora	Ž5034	Connection to Ravna Gora and Kupjak[20]
	42.0	4	Delnice	D3	Connection to Delnice, Mrkopalj, and Risnjak National Park[20]
	45.2				Golubinjak Viaduct
	50.6	5	Vrata	Ž5068	Connection to Vrata, Fužine, and Lake Bajer[20]
	51.9				Bajer Bridge
	56.3	P	Lepenica rest area		Accessible to southbound traffic only
	57.6				Tuhobić Tunnel
	59.0	P	Tuhobić rest area		Accessible to northbound traffic only
	59.8				Hreljin Viaduct
	61.7	6	Oštrovica	D501	Connection to Crikvenica, Kraljevica, and Krk island via Križišće[20]
	70.1	⊚	Grobnik toll plaza		
	70.4	7a	Kikovica	D3	Connection to Automotodrom Grobnik[20]
	74.5	P	Cernik-Čavle rest area		Accessible to northbound traffic only
	75.6	7b	Čavle	D40	Connection to Čavle, Kukuljanovo, and Bakar[20]
	80.2	8	Orehovica	A7 E61 E65 Ž5054	Connection to the A7 motorway in Orehovica interchange and to Rijeka via Rijeka bypass section of the A7[2] Connection to Rijeka via Ž5054 (exit only)[55] The southern terminus of European route E65 concurrency The southern terminus of the motorway
1.000 mi = 1.609 km; 1.000 km = 0.621 mi					
Concurrency terminus • Closed/former • Incomplete access • Unopened					

See also

* Wikipedia book: A6 motorway and its major structures (2010)
* International E-road network
* Transport in Croatia

References

[1] "Overview of motorways and semi-motorways" (http://www.huka.hr/Motorways-network/). HUKA. . Retrieved September 8, 2010.

[2] "Pravilnik o označavanju autocesta, njihove stacionaže, brojeva izlaza i prometnih čvorišta te naziva izlaza, prometnih čvorišta i odmorišta [Regulation on motorway markings, chainage, interchange/exit/rest area numbers and names]" (http://narodne-novine.nn.hr/clanci/ sluzbeni/305463.html) (in Croatian). *Narodne novine*. May 6, 2003. . Retrieved September 6, 2010.

[3] "Transport : launch of the Italy-Turkey pan-European Corridor through Albania, Bulgaria, Former Yugoslav Republic of Macedonia and Greece" (http://europa.eu/rapid/pressReleasesAction.do?reference=IP/02/1275&format=HTML&aged=0&language=EN;& guiLanguage=en). European Union. September 9, 2002. . Retrieved September 6, 2010.

[4] Jelena Lončar (December 14, 2007). "Međuovisnost prometa i turizma u Hrvatskoj [Interdependency of transport and tourism in Croatia]" (http://www.geografija.hr/clanci/1301/meduovisnost-prometa-i-turizma-u-hrvatskoj) (in Croatian). geografija.hr. . Retrieved September 6, 2010.

[5] "Proširenje lučkih kapaciteta u Rijeci [Expansion of Port of Rijeka facilities]" (http://www.hrt.hr/index.php?id=48& tx_ttnews[tt_news]=89102&cHash=c51ec7cf44) (in Croatian). Croatian Radiotelevision. October 3, 2010. . Retrieved October 8, 2010.

[6] Darko Pajić (July 3, 2010). "Četiri poslovne zone za 40 milijuna tona tereta riječke luke [Four business zones for 40 million tons of cargo handled by Port of Rijeka]" (http://www.novilist.hr/2010/07/06/cetiri-poslovne-zone-za-40-milij.aspx) (in Croatian). *Novi list*. . Retrieved October 8, 2010.

[7] "Realizacija prometnog čvora Rijeka kao pretpostavka gospodarskog razvoja županije [Execution of Rijeka transport node as a precondition of economic development of the county]" (http://www.pgz.hr/PDF/Realizacija prometnog cvora Rijeka - Zlatko Komadina.pdf) (in Croatian) (PDF). Primorje-Gorski Kotar County. September 2, 2003. . Retrieved October 8, 2010.

[8] "Croatian Motorways (pp. 322–365)" (http://www.hac.hr/files/file/brosure/monografija/virtualMagazine.html). Hrvatske autoceste. . Retrieved October 7, 2010.

[9] "Tolls - price list" (http://www.arz.hr/index.php?page=4&sub=1&&lng=2&NP_Ulaz=105). Autocesta Rijeka–Zagreb. . Retrieved October 10, 2010.

[10] Igor Žic (February 13, 2008). "Od Lujzijane do autoceste [From Louisiana to the motorway]" (http://www.ppv.pgz.hr/dokumenti/ ZMIGAVAC_7_1.pdf) (in Croatian) (PDF). *Žmigavac* (Primorje-Gorski Kotar County). . Retrieved October 7, 2010.

[11] Jakša Miličić (2004). "Autocesta Split - Zagreb [Autocesta Split - Zagreb]" (http://www.matica.hr/HRRevija/revija2004_4_n.nsf/ AllWebDocs/Milicic) (in Croatian). *Hrvatska revija*. Matica hrvatska. . Retrieved May 16, 2010.

[12] "Otvoren puni profil autoceste Rijeka-Zagreb [Full cross-section of Rijeka-Zagreb motorway opens]" (http://www.mmpi.hr/default. aspx?id=5251) (in Croatian). Ministry of Sea, Transport and Infrastructure. October 22, 2008. . Retrieved October 7, 2010.

[13] "Odluka o osnivanju dioničkog društva Autocesta Rijeka - Zagreb d.d. i dodjeli koncesije za građenje i gospodarenje autocestom Rijeka - Zagreb [Decision on founding of Rijeka - Zagreb Motorway joint stock company and granting of concession regulating construction and management of Rijeka - Zagreb motorway]" (http://narodne-novine.nn.hr/clanci/sluzbeni/1997_12_139_1994.html) (in Croatian). *Narodne Novine*. December 11, 1997. . Retrieved September 6, 2010.

[14] "Zakon o javnim cestama [Public Roads Act]" (http://narodne-novine.nn.hr/clanci/sluzbeni/2004_12_180_3130.html) (in Croatian). *Narodne Novine*. December 14, 2004. . Retrieved September 6, 2010.

[15] Google, Inc. *Google Maps – A6 (Croatia)* (http://maps.google.com/maps?f=d&source=s_d&saddr=A1&daddr=45.4028947,15. 2413832+to:45.3909341,15.1985347+to:45.3853659,15.182387+to:E65& geocode=FeThtAIdMvPoAA;FQ7LtAIdp5DoACmXq8sc225kRzFgP5cM4anoTw;FVactAIdRunnACnXR6R9eWVkRzHNuGjEs9To5g;FZWGtAIdM6rnACnX hl=en&mra=dvme&mrcr=0&mrsp=1&sz=11&via=1,2,3&sll=45.385913,15.238037&sspn=0.312979,0.727158&ie=UTF8&ll=45. 341528,14.94278&spn=0.607142,1.454315&t=h&z=10) (Map). Cartography by Google, Inc. . Retrieved September 17, 2010.

[16] "European Agreement on main international traffic arteries (AGR) (with annexes and list of roads). Concluded at Geneva on 15 November 1975" (http://treaties.un.org/doc/Publication/UNTS/Volume 1302/volume-1302-I-21618-English.pdf) (PDF). United Nations. . Retrieved July 29, 2011.

[17] Crnjak, Mario; Puž, Goran (November 2007) (PDF). *Kapitalna prometna infrastruktura [Capital transport infrastructure]* (http://www. mmpi.hr/UserDocsImages/Kapitalna_prometna_infrastruktura.pdf). Hrvatske autoceste. pp. 37–39. ISBN 9789537491022. . Retrieved October 11, 2010.

[18] Tanja Poletan Jugović (April 11, 2006). "The integration of the Republic of Croatia into the Pan-European transport corridor network" (http://hrcak.srce.hr/file/6570). *Pomorstvo* (University of Rijeka, Faculty of Maritime Studies) **20** (1): 49–65. . Retrieved October 14, 2010.

[19] "Rest areas - types and facilities" (http://www.arz.hr/index.php?page=6&sub=2&lng=2). Autocesta Rijeka–Zagreb. . Retrieved September 7, 2010.

[20] "Odluka o razvrstavanju javnih cesta u državne ceste, županijske ceste i lokalne ceste [Decision on categorisation of public roads as state roads, county roads and local roads]" (http://narodne-novine.nn.hr/clanci/sluzbeni/2010_02_17_410.html) (in Croatian). *Narodne Novine*. February 17, 2010. . Retrieved September 6, 2010.

[21] (PDF) *Croatian Motorways* (http://www.hac.hr/files/file/brosure/monografija/virtualMagazine.html). Hrvatske autoceste. 2007. pp. 130–133. ISBN 9789537491. . Retrieved September 5, 2010.

[22] "Autocesta Rijeka–Zagreb: Zaštita voda u funkciji održivog razvoja Autoceste Rijeka–Zagreb [Rijeka–Zagreb Motorway: Water protection for sustainable development of Rijeka–Zagreb Motorway]" (http://www.huka.hr/objekti/publikacije/hr/2009_16.pdf) (in Croatian). *HUKA bilten* (HUKA) (16): 4. March 2009. . Retrieved October 10, 2010.

[23] "Rijeka - Zagreb Motorway - Information" (http://www.arz.hr/index.php?page=2&sub=2&lng=2). Autocesta Rijeka–Zagreb. June 9, 2010. .

[24] Damir Herceg (October 6, 2006). "Kroz špilju u tunelu Vrata gradit će se most! [A bridge shall be built across a cavern in Vrata Tunnel]" (http://www.vjesnik.hr/pdf/2006\10\06\06A6.PDF) (in Croatian) (PDF). *Vjesnik*. . Retrieved October 12, 2010.

[25] Branko Nadilo (December 20, 2006). "Gradnja do punog profila na autocesti Rijeka-Zagreb [Expansion of Rijeka-Zagreb motorway to full motorway]" (http://www.casopis-gradjevinar.hr/dokumenti/200611/5.pdf) (in Croatian) (PDF). *Građevinar*. . Retrieved October 7, 2010.

[26] Jakov Prkić (March 13, 2009). "Utroba Biokova puna jama - na trasi Sv. Ilije dvije "bezdanke" [Multitude of pits in Biokovo - two abysses found in Sveti Ilija Tunnel route]" (http://www.slobodnadalmacija.hr/Split-županija/tabid/76/articleType/ArticleView/articleId/45858/Default.aspx) (in Croatian). *Slobodna Dalmacija*. . Retrieved October 12, 2010.

[27] "Statistički podaci [Electronic toll collection available on HAC, ARZ and BINA Istra motorways]" (http://www.huka.hr/Cestarine/ENC/) (in Croatian). HUKA. . Retrieved August 28, 2010.

[28] "Statistički podaci [Statistical data]" (http://www.huka.hr/objekti/publikacije/hr/2010_19.pdf) (in Croatian) (PDF). *HUKA bilten* (HUKA) (19): 8. July 2010. . Retrieved September 6, 2010.

[29] Zdravko Duplančić, Ivan Banjad (September 27, 2009). "Opravdanost izgradnje šesterotračne autoceste Zagreb - Karlovac [Feasibility of construction of six traffic lanes of Zagreb - Karlovac motorway]" (http://www.gradimo.hr/Opravdanost-izgradnje-sesterotracne-autoceste-Zagreb---Karlovac/hr-HR/12991.aspx) (in Croatian). Gradimo. . Retrieved October 7, 2010.

[30] "Autocesta A1: Zagreb–Split [A1 motorway: Zagreb–Split]" (http://www.hac.hr/files/file/brosure/prezentacija-HR.pdf) (in Croatian) (PDF). Hrvatske autoceste. July 18, 2005. . Retrieved July 29, 2011.

[31] "Autocesta A6, Rijeka-Zagreb otvorena i puštena u promet u punom profilu [A6 Rijeka-Zagreb motorway open for traffic as a full motorway]" (http://www.huka.hr/v2/objekti/publikacije/hr/2008_15.pdf) (in Croatian) (PDF). *HUKA bilten* (HUKA) (15): 1. November 2008. . Retrieved October 7, 2010.

[32] Alen Legović (April 21, 2009). "Ponovo najbolje ocjene za hrvatske tunele [Again, top score awarded to Croatian tunnels]" (http://www.dw-world.de/dw/article/0,,4195171,00.html) (in Croatian). Deutsche Welle. . Retrieved October 7, 2010.

[33] "Rijeka–Zagreb motorway, section: Oštrovica–Vrata" (http://www.arz.hr/?page=3&sub=4&lng=2). Autocesta Rijeka–Zagreb. . Retrieved October 7, 2010.

[34] Damir Herceg (August 10, 2007). "Tunel Tuhobić probijen dva mjeseca prije roka [Tuhobić Tunnel excavated two months ahead of schedule]" (http://www.vjesnik.hr/Pdf/2007\08\10\04A4.PDF) (in Croatian) (PDF). *Vjesnik*. . Retrieved October 7, 2010.

[35] "EuroTAP testiranje tunela: Sveti Rok dio europske elite [EuroTAP tunnel test: Sveti Rok Tunnel as a part of European elite]" (http://www.hak.hr/novosti/vijesti-iz-hak-a/eurotap-testiranje-tunela-sveti-rok-dio-europske-elite-(1).aspx). Hrvatski Autoklub. July 15, 2010. . Retrieved June 8, 2010.

[36] "Rijeka–Zagreb motorway, section: Ravna Gora–Vrbovsko" (http://www.arz.hr/?page=3&sub=7&lng=2). Autocesta Rijeka–Zagreb. . Retrieved October 7, 2010.

[37] Tihomir Ponoš (June 27, 2004). "Naša ideja je bila izgraditi ono što samo zvali 'hrvatski križ' [Our idea was to build what we called 'Croatian cross']" (http://www.vjesnik.hr/Pdf/2004\06\27\09A9.PDF) (in Croatian) (PDF). *Vjesnik*. . Retrieved October 7, 2010.

[38] "Priča duga 35 godina: Rijeka i Zagreb spojeni autocestom [35 years long story: Rijeka and Zagreb connected by a motorway]" (http://www.huka.hr/v2/objekti/publikacije/hr/2004_02.pdf) (in Croatian) (PDF). *HUKA bilten* (HUKA) (2): 1. September 2004. . Retrieved October 7, 2010.

[39] Branko Nadilo (August 11, 2003). "Posljednje dionice na autocesti Rijeka - Zagreb [The last sections of Rijeka - Zagreb motorway]" (http://www.casopis-gradjevinar.hr/dokumenti/200305/6.pdf) (in Croatian) (PDF). *Građevinar*. . Retrieved October 8, 2010.

[40] Dalibor Klobučar (December 23, 2005). "Mreža autocesta ispletena do pola: Kako skupiti još 19,5 milijardi kuna [Motorway network only half complete - How to raise additional 19.5 billion Kuna]" (http://www.poslovni.hr/vijesti/mreza-autocesta-ispletena-do-pola-kako-skupiti-jos-195-milijardi-kuna-1622.aspx) (in Croatian). *Poslovni dnevnik*. . Retrieved October 8, 2010.

[41] "Otvara se dionica Vrata-Delnice-Ravna Gora na autocesti Rijeka-Zagreb [Vrata-Delnice-Ravna Gora section of Rijeka-Zagreb Motorway opens]" (http://www.liderpress.hr/Default.aspx?sid=48087) (in Croatian). *Lider Press*. June 26, 2008. . Retrieved October 7, 2010.

[42] "Otvoren most 'Mura' i dionica Goričan - Letenye [Mura Bridge and Goričan - Letenye section open]" (http://dnevnik.hr/vijesti/hrvatska/otvoren-most-mura-i-dionica-gorican-letenye.html) (in Croatian). Nova TV. October 22, 2008. . Retrieved October 7, 2010.

[43] Tomislav Grdić, Damir Herceg (October 23, 2008). "Otvoren koridor Rijeka-Zagreb-Budimpešta [Rijeka-Zagreb-Budapest corridor opens]" (http://www.vjesnik.hr/Html/2008/10/23/Clanak.asp?r=unu&c=16) (in Croatian). *Vjesnik*. . Retrieved October 7, 2010.

[44] "Traffic counting on the roadways of Croatia in 2009 - digest" (http://www.hrvatske-ceste.hr/WEB - Legislativa/brojenje-prometa/CroDig2009.pdf) (PDF). Hrvatske ceste. May 1, 2010. . Retrieved September 6, 2010.

[45] "Basic types and offer of roadside service facilities" (http://www.hac.hr/en/motorways/rest-areas/basic-types/). Hrvatske autoceste. . Retrieved July 29, 2011.

[46] Darko Pajić (June 4, 2010). "Ina "prosula" 500 tisuća eura u Gorskom kotaru [INA "spills" 500 thousand euro in Gorski Kotar]" (http://www.novilist.hr/2010/06/06/ina--BBprosula-AB-500-tisuca-eur.aspx) (in Croatian). *Novi list*. . Retrieved September 26, 2010.

[47] "Rest areas - Ravna Gora" (http://www.arz.hr/index.php?page=6&sub=8&lng=2). Autocesta Rijeka–Zagreb. . Retrieved October 7, 2010.

[48] "Lista benzinskih postaja (br. 11) [Filling station list (item 11)]" (http://www.tifon.hr/default.asp?ru=99&akcija=) (in Croatian). Tifon. . Retrieved October 7, 2010.

[49] "Marché Ravna Gora" (http://marche.moevenpick.com/#/restaurant/empty/138/empty/en/). Marché. . Retrieved October 7, 2010.

[50] "Rest areas - Lepenica" (http://www.arz.hr/index.php?page=6&sub=9&lng=2). Autocesta Rijeka–Zagreb. . Retrieved October 7, 2010.

[51] "Petrol Station Search - A1 motorway" (http://www.ina.hr/default.aspx?id=475). INA. . Retrieved September 26, 2010.

[52] "Otvoren prateći uslužni objekt Tuhobić-jug [Tuhobić-jug rest area opens]" (http://www.arz.hr/novosti/puotuh.php?lng=1) (in Croatian). Autocesta Rijeka–Zagreb. . Retrieved November 21, 2010.

[53] "Rest areas - Tuhobić" (http://www.arz.hr/index.php?page=6&sub=14&lng=2). Autocesta Rijeka–Zagreb. . Retrieved November 21, 2010.

[54] "Rest areas - Cernik-Čavle" (http://www.arz.hr/index.php?page=6&sub=10&lng=2). Autocesta Rijeka–Zagreb. . Retrieved October 7, 2010.

[55] "Rekonstrukcija i dogradnja čvora Orehovica na državnom cestovnom pravcu D3 [Reconstruction and expansion of Orehovica interchange on D3 state road route]" (http://www.rijeka.hr/RekonstrukcijaIDogradnja) (in Croatian). City of Rijeka. . Retrieved October 12, 2010.

External links

- Autocesta Rijeka–Zagreb web cameras (http://www.arz.hr/index.php?page=5&sub=7&lng=2&cam=1)
- A6 motorway (http://en.structurae.de/structures/data/index.cfm?ID=p0000366) at *Structurae*
- Exit list of A6 (http://www.motorways-exitlists.com/europe/hr/a6.htm)

Gorski_kotar

Gorski kotar (pronounced [ɡȏrski̇̀ kôtaːr]; English: *Mountain District*) is the mountainous region in Croatia between Karlovac and Rijeka. Together with Lika and the Ogulin-Plaški valley it forms Mountainous Croatia. Because 63% of its surface is forested it is popularly called *the green lungs of Croatia* or *Croatian Switzerland*. Through the region passes the European route E65, which connects Budapest and Zagreb with the Adriatic Port of Rijeka.

Geography

The region is divided between Primorje-Gorski Kotar county and Karlovac county. The majority of the region lies in Primorje-Gorski kotar county including the cities of Delnice, Čabar, Vrbovsko; and the municipalities of Mrkopalj, Ravna Gora, Skrad, Brod na Kupi, Fužine and Lokve. The part of the region that is in Karlovac county contains the Municipality of Bosiljevo and part of the City of Ogulin. With a population of 4454, Delnice is the largest city of the region and its center. Other centers with populations of more than 1,000 are Vrbovsko (1,900) and Ravna Gora (1,900). Begovo Razdolje, the highest town in Croatia, is located in Gorski kotar at an altitude of 1076 m.

Lokve lake at Gorski kotar, Mt. Risnjak in the distance.

The population density of Gorski kotar is low, but the highest in Mountainous Croatia. As of 2001, about 28,000 people were living in an area of about 1300 square kilometers, which is about 22 people per square kilometer.

Linguistically, the region is very diverse; in a relatively small area, all three Croatian dialects can be heard. Also, in some parts of Gorski kotar there is a considerable Serb community, especially in the city of Vrbovsko where they constitute 36% of the population.

Geomorphically Gorski kotar is on the karstic plateau about 35 km and has an average altitude of 800 m. The highest point is Bjelolasica at 1534 m followed by Risnjak at 1528 m. The plateau is a climatic barrier between the littoral and continental parts of the country. The border with the Kvarner region is defined by the divide between the drainage basins of the Black Sea and of the Adriatic Sea. Its southern border with Lika is not clearly defined but most scholars consider it to be the Jasenak – Novi Vinodolski road and people of Jasenak consider themselves to be somewhere between Gorski kotar and Lika. To the north, the Kupa river is the border between Gorski kotar and Slovenia's Bela Krajina region.

Tourism

Popular tourist destinations in Gorski Kotar include Lake Lokvarsko in the village of Lokve. This is the seventh largest lake in the whole of Croatia measuring over two square kilometres. Nature walks are possible right around the lake. You can also swim and canoe in the summer and fish all year round. A world record 25 kg trout was once caught in this very lake.

Other attractions in Lokve include Golubinjak Nature Forest Park and Lokvarka Caves.

Gorski Kotar's main town is called Delnice. It's approximately 5 miles from Lokve. Delnice has many shops, bars and restaurants as well as train & bus connections to numerous local and national destinations.

The nearest airport to Gorski Kotar is Rijeka Airport (RJK) located on the island of Krk about half an hour's drive away. Zagreb Airport is also a convenient gateway to Gorski Kotar just over an hour's drive away. Slightly further afield are Pula Airport and Ljubljana Airport (each are about 2 hours away) and Trieste Airport in Italy which is about two and a half hours drive away.

Skiing is popular in several places in Gorski Kotar including the resort village of Mrkopalj. Platak is also popular for skiing although technically it falls into the neighbouring region of Kvarner not Gorski Kotar. Platak boasts some truly stunning views of the Adriatic Sea from the pistes.

History

The first known inhabitants of the Gorski kotar was the Illyrian tribe of the Iapodes, who lived in the area from the 9th century BC on. They were later subjugated by the Romans, who built lines of fortification from Grobnik to Prezid. In the 6th century, the Slavs began to settle the area, but hereafter the history of Gorski kotar is obscure until the 12th century when the noble family Frankopans began to rule much of Gorski kotar. The Frankopans initiated the first wave of settlement in the 14th century, first colonizing the eastern part of the Gorski kotar, making their stronghold at Bosiljevo.

In the 15th century, due to the Ottoman intrusions, the geopolitical significance of the Gorski kotar increased. This led to a new wave of settlement and the creation of defensive fortifications, which in turn led to the development of more important towns in the region. After a short period of insecurity, settlement renewed at the end of 16th century, when many Ottoman exiles and refugees (mostly Shtokavian-speaking, and many Orthodox Christians) came to this region. They settled along the border with the Ottoman empire: in Gomirje, Vrbovsko, Dobra, Moravice, Stari Laz, Sušica, Mrkopalj and Lič. From the mid-17th century until the beginning of the 18th century, the most developed part of the region was the area around Čabar, where the Zrinskis had control overiron mines and metallurgic manufacturing. Čabar and its surroundings were settled by a Slovene population from Carniola and Chakavian from Kvarner. After the failed Zrinski-Frankopan plot their properties were confiscated and shared among many other nobles, whereupon the local population was exploited with increased severity.

The most intense period of settlement began in the 18th century with the 1732 opening of the Karolina road, which linked Karlovac and Bakar. Most immigrants were from Kvarner but Czechs, Slovenes and descendants of Ottoman exiles also came. After the road opened, economic activity in the region flourished, especially *kirijašenje* - the transportation of goods from the interior to the Adriatic. The most developed center in this period was Ravna Gora. In 1777 by decree of Maria Theresia, all of the Gorski kotar was incorporated into the county of Severin.

During the Napoleonic wars, the Gorski kotar was part of France's Illyrian Provinces. The French built a newer, wider road Lujziana, which connected Karlovac and Rijeka and led to more development of Gorski kotar, and Delnice became the most developed center in the region. The road is presently still in use.

After the war, the Gorski kotar was again under the Habsburg monarchy, and in 1873, the first railway in the region was built. However, this decreased dependence on *kirijašenje* and an economic crisis ensued that forced many to leave the region. In 1886, a new administrative division was made and all of the Gorski kotar was incorporated into Rijeka county.

After the First World war, the Gorski kotar was part of the Kingdom of Serbs, Croats and Slovenes.

On 21 July 1921 Alija Alijagić, a member of the communist organization *Crvena Pravda*, shot the Minister of the Interior Milorad Drašković in Delnice. Although Milorad Drašković was a staunch anti-communist, and enacted several pieces of anti-communist legislation, the Communist Party of Yugoslavia condemned the act. Nevertheless, this inspired the King to make a 'law concerning protection of the state' that made the communist party illegal.

During the Second World War, the Gorski kotar was divided between Italy and the Independent State of Croatia. Citizens of the Gorski kotar participated greatly to the anti-fascist struggle, and several popular post-war TV-series were made about Gorski kotar resistance efforts, *Kapelski kresovi* being one.

In the 1990s, construction of the Zagreb − Rijeka motorway was resumed, part of it having already been built in 1973 (Zagreb − Karlovac and Rijeka − Oštrovica), and in 1997 Delnice was connected to Rijeka. In 2004 the motorway was finished by connecting the two directions of motorway in Vrbovsko. The motorway is now known as A6 and it has also been upgraded to the full two-lane format in 2008.

External links

- Gorski kotar [1] English language home page.
- Gorski Kotar images [2]
- Gorski Kotar images [3]

References

[1] http://www.rasip.fer.hr/gorski_kotar
[2] http://www.sobol.hr/fotogalerija-gorski-kotar2.htm
[3] http://www.gorskikotar.com/download.php

Dobra_(river)

The **Dobra** is a river located mostly in the Karlovac County in the Republic of Croatia. It is 104 kilometres (65 mi) long and its basin covers an area of 900 square kilometres (350 sq mi).[1]

Dobra rises in Gorski Kotar near Skrad and Ravna Gora, where it flows first to the north and then turns to the east. It flows past Vrbovsko, to the southeast into the city of Ogulin, where it becomes an underground stream. It takes a sharp northward turn and rises back to the surface north of Ogulin. It continues to the northeast, past the Lešće spa and a hydroelectric plant (built and in test operation as of 2010), running in parallel to the Kupa and Mrežnica, and finally flows into the Kupa north of Karlovac.

North of Ogulin, near Gojak, the water of Dobra is harvested for the Gojak Hydroelectric Power Plant, a hydroelectric power plant built to utilize the rivers Dobra and Mrežnica.

Two motorway bridges have been built over the Dobra: the Dobra Bridge (A1) and the Dobra Bridge (A6).

References

[1] "Geographical and meteorological data" (http://www.dzs.hr/Hrv_Eng/ljetopis/2009/PDF/01-bind.pdf). *Statistical Yearbook.* Croatian Bureau of Statistics. 2009. . Retrieved 2011-07-10.

Ticket_system

For the IT term, see issue tracking system.

A **ticket system** toll road (also known as **closed toll collection system**, as opposed to a flat-rate toll road, is utilized by some state toll road or highway agencies that allows a motorist to pay a toll rate based on the number of miles traveled from their origin to their destination exit.

The correct toll rate per user is easily determined by requiring all users to take a ticket from a machine or from an attendant when entering the system. The ticket prominently displays the location (or exit number) from which it was dispensed and a precomputed chart of toll rates with a list of all exits on one axis and various sizes of vehicles on the other axis. Upon arrival at the toll booth at the destination exit, the motorist

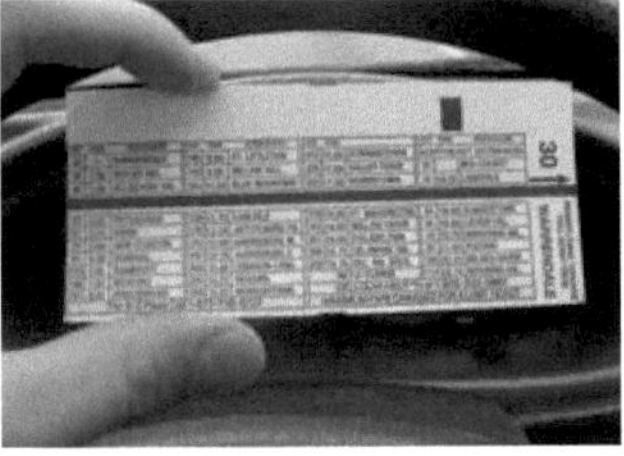

A toll ticket used on the Pennsylvania Turnpike

presents the ticket to the toll collector, who matches the axis for that exit against the axis for the motorist's vehicle and demands the correct toll.

First employed on the Pennsylvania Turnpike when it opened in 1940, it has been utilized on lengthy toll highways in which the exits are spread out over a distance on an average of 7 to 10 miles (11 to 16 km) per exit. Flat-rate highways, on the other hand, have mainline toll booths placed at equal distances on the highway, with ramps, depending on the direction of travel, having either coin or token-drop baskets or toll barriers or no barriers at all.

Highways that use the ticket system

* Massachusetts Turnpike — between New York State Line and Exits-14/15
* New Jersey Turnpike — entire length including Pearl Harbor and Newark Bay Extensions, except for that part of the eastern spur north of exit 16; also note that the northernmost service area (Vince Lombardi) is outside the ticket system (and because it serves both north- and southbound traffic, it is possible to access that service area toll-free).
* Pennsylvania Turnpike — between Warrendale Main Line Barrier and Delaware River Bridge interchange on East-West Main Line and between Mid-County and Wyoming Valley interchanges on Northeast Extension
* New York State Thruway — between New York Route 17 and Buffalo (including Berkshire Section) and between Buffalo and Pennsylvania State Line (Erie Section).
* Kansas Turnpike — entire length
* Ohio Turnpike — entire length
* Turner Turnpike and Will Rogers Turnpike in Oklahoma — both use a modified ticket-based toll collection system that places only one mainline toll plaza on the highway, roughly halfway through the length of the road. Under this system, traffic exiting before reaching the mainline toll plaza pays at the exit. Also, traffic entering at said interchange heading away from the mainline toll plaza pays before entering the highway, but traffic entering heading toward the mainline toll plaza receives a ticket. Traffic heading away from the mainline toll plaza that exits before reaching the end of the toll road turns in their receipt they received when paying their toll and receives a refund for the unused portion of the toll road.
* Indiana Toll Road — between Portage toll barrier and the Eastpoint Barrier at the Ohio state line.
* Florida's Turnpike — between Lantana and Kissimmee.
* Almost all toll highways in China, Croatia, Serbia, France, Italy, and Indonesia.

Autocesta_Rijeka_–_Zagreb

Type	State-owned joint-stock company
Industry	Road transport
Founded	1998
Headquarters	Zagreb, Croatia
Key people	Miro Škrgatić
Website	www.arz.hr [1]

Autocesta Rijeka - Zagreb (English: *Rijeka - Zagreb Motorway*) is a Croatian state-owned joint-stock company founded pursuant to decision of the government of the Republic of Croatia of December 11, 1997, to facilitate construction and subsequent management of a motorway between Rijeka and Zagreb.[2] The company started operating on March 15, 1998. The company issued 21,520 shares, nominally valued at 100,000.00 Croatian kuna each. All company stocks are owned by the Republic of Croatia.[3]

The company was granted the motorway management concession for a period of 28 years, which has subsequently been expanded to include not only the A6 motorway and some sections of the A1 motorway, but also the A7 motorway and the Krk Bridge on the D102 state road.[4] [5] The expansion of the concession also included extension of the concession period to 32 years and 11 months.

The company currently manages or develops the following routes:

Number	Control cities *(or other appropriate route description)*
A1	Bosiljevo (A6) - Lučko (A3)
A6	Bosiljevo (A6) - Rijeka
A7	Rupa border checkpoint - Rijeka (A6) - Sveti Kuzam - Križišće
D102	Krk Bridge *(maintenance and bridge toll collection)*

The company is currently managed by Miro Škrgatić (chairman of the managing board) and a five-member supervisory board.[6]

See also

- Highways in Croatia
- Hrvatske autoceste
- Hrvatske ceste

References

[1] http://www.arz.hr/

[2] "Decision on founding of Rijeka - Zagreb Motorway joint stock company and granting of concession regulating construction and management of Rijeka - Zagreb motorway" (http://narodne-novine.nn.hr/clanci/sluzbeni/1997_12_139_1994.html) (in Croatian). *Narodne Novine*. December 11, 1997. .

[3] "ARZ general info" (http://www.arz.hr/?page=2&lng=1) (in Croatian). *Autocesta Rijeka - Zagreb*. August 20, 2010. .

[4] "Decision on amendments and additions to decision on founding of Rijeka - Zagreb Motorway joint stock company and granting of concession regulating construction and management of Rijeka - Zagreb motorway" (http://narodne-novine.nn.hr/clanci/sluzbeni/2007_08_82_2585.html) (in Croatian). *Narodne Novine*. August 2, 2007. .

[5] "ARZ concession takes over Rijeka node" (http://www.liderpress.hr/Default.aspx?sid=25291) (in Croatian). *Lider Press*. August 30, 2007. .

[6] "ARZ structure" (http://www.arz.hr/?page=2&sub=2&lng=1) (in Croatian). *Autocesta Rijeka - Zagreb*. August 20, 2010. .

External links

- Official website (http://http://www.arz.hr/)

Adriatic_Sea

Adriatic Sea	
[[Image:Adriatic Sea map.png	Adriatic Sea} - Map of the Adriatic Sea]] Map of the Adriatic Sea
Location	Europe
Coordinates	43°N 15°E
Catchment area	235000 km^2 (91000 sq mi)
Max length	800 km (500 mi)
Max width	200 km (120 mi)
Surface area	138600 km^2 (53500 sq mi)
Average depth	252 m (827 ft)
Max depth	1233 m (4045 ft)
Water volume	35000 km^3 (8400 cu mi)
Salinity	38-39 PSU
Max temperature	24 °C (75 °F)
Min temperature	9 °C (48 °F)

The **Adriatic Sea** (◀ /ˌeɪdriˈætɪk/) is a body of water separating the Apennine Peninsula from the Balkan peninsula, and the Apennine Mountains from that of the Dinaric Alps and adjacent ranges. The Adriatic Sea is the northernmost arm of the Mediterranean Sea extending from the Strait of Otranto, where it connects to the Ionian Sea, to the Northwest and the Po Valley. Its coasts belong to Italy, Slovenia, Croatia, Bosnia and Herzegovina, Montenegro and Albania. The Adriatic contains more than a thousand islands, largely located along its Eastern coast. The Adriatic is divided into three basins—the Northern being the shallowest and the Southern being the deepest, where the maximum depth of 1233 metres (4045 feet) is recorded. Otranto Sill is located at the border of the Adriatic and Ionian seas. Prevailing currents flow in counterclockwise direction from the Strait of Otranto, along the Eastern coast and back to the strait along the Western, Italian coast. Tidal movement is slight, although larger amplitudes are known to occur occasionally. Salinity of the Adriatic is lower compared to the Mediterranean as the former collects a third of freshwater flowing into the latter, acting as a dilution basin. Surface water temperature generally ranges from 24 °C (75 °F) in summer to 12 °C (54 °F) in winter, significantly moderating climate of the Adriatic basin. Shores of the Adriatic are populated by more than 3.5 million people, and the largest cities are Bari, Venice, Trieste and Split.

The Adriatic Sea sits on Apulian or Adriatic Microplate which separated from the African Plate in Mesozoic. Movement of the plate contributed to Alpine orogeny and uplift of the Apennines. All types of sediment are found in the Adriatic, with bulk of the material transported by the Po and other rivers on the West coast. The Western coast is alluvial or terraced, while the Eastern coast is well indented with pronounced karstification. There are 15 marine protected areas in the Adriatic designed to protect the karst habitats and biodiversity of the sea. The sea is abundant in flora and fauna—more than 7,000 species are identified as native to the Adriatic, including endemic, rare and threatened ones.

The earliest settlements on the Adriatic shores were Etruscan, Illyrian, and Greek. By the 2nd century BC, the shores were under full control of the Rome. In the Middle Ages, the Adriatic shores and the sea itself was controlled, to a varying extent, by a series of states—most notably the Byzantine Empire, the Republic of Venice, the Habsburg Monarchy and the Ottoman Empire. The Napoleonic Wars resulted in French control of the coasts and British effort to counter the French in the area, ultimately securing most of the Eastern Adriatic and the Po Valley for Austria. Following unification, Italy started eastward expansion which would last until the 20th century. Following the World War I and collapse of Austria-Hungary and the Ottoman Empire, control of the Eastern coast passed to Yugoslavia and Albania. The former disintegrated in 1990s, resulting in four new states on the Adriatic shores.

The Adriatic Sea is significant for economies of the countries found along its coasts, especially in terms of fisheries and tourism. The Adriatic Croatia recently records comparably greater growth relative to the rest of the Adriatic basin. Maritime transport is also a significant branch of economy in the area—there are 19 major seaports in the Adriatic handling more than a million tonnes of cargo per year. The largest Adriatic seaport by annual cargo turnover is the Port of Trieste, while the Port of Split is the largest Adriatic seaport in terms of number of passengers served per year. Italy and former Yugoslavia defined their maritime boundaries by 1975 and the boundary is recognized by Yugoslavia's successor states, but maritime boundaries between Slovene, Croatian, Bosnia and Herzegovinian and Montenegrin waters are disputed. Italy and Albania defined their maritime boundary in 1992.

Geography

Adriatic Sea is bordered by the Apennine peninsula in the southwest, Italian regions of Veneto and Friuli-Venezia Giulia in the northwest, and in the northeast by Slovenia, Croatia, Bosnia and Herzegovina, Montenegro and Albania—the Balkan peninsula. In the southeast, the Adriatic Sea connects to the Ionian Sea at the 72-kilometre (45 mi) wide Strait of Otranto.[1] The International Hydrographic Organization (IHO) defines the boundary between the Adriatic and the Ionian seas as a line running from the mouth of the Butrinto River (39°44'N) in Albania to Cape Karagol in Corfu, through this island to Cape Kephali (these two capes are in lat. 39°45'N) and on to Cape Santa Maria di

Bay of Kotor, a ria in the Southern Adriatic

Leuca.[2] It extends 800 kilometres (500 miles) from the northwest to the southeast and it is 200 kilometres (120 miles) wide. It covers 138600 square kilometres (53500 square miles) and comprises volume of 35000 cubic kilometres (8400 cubic miles). The Adriatic extends northwest from 40° to 45°47' North, representing the northernmost portion of the Mediterranean.[1] Adriatic Sea drainage basin encompasses 235000 square kilometres (91000 square miles), yielding land to sea ratio of 1.8. Mean elevation of the drainage basin is 782 metres (2566 feet) above sea level, while its mean slope is 12.1°.[3] Major rivers discharging into the Adriatic are Po, Soča, Krka, Neretva, Drin, Bojana and Vjosë rivers.[4] In late 19th century, Austria-Hungary established a geodetic network, whose elevation benchmark was determined on the basis of oscillations of Adriatic Sea level at Sartorio pier in Trieste. The benchmark was subsequently retained by Austria, adopted by Yugoslavia and retained by states which emerged after its dissolution.[5] [6]

Length of coastlines of the Adriatic Sea [7] [8]				
Country	Mainland	Islands	Total	Coastal front*
Croatia	1,777.3	4,058	5,835.3	526
Italy	1,249	23	1,272	926
Albania	396	10	406	265
Montenegro	249	11	260	92
Slovenia	46.6	N/A	46.6	17
Bosnia and Herzegovina	21.2	N/A	21.2	10.5
*The distance between the extreme points of each state's coastline				

The Adriatic Sea contains well more than a thousand islands and islets. By far, the most of them are located along the Eastern coast of the Adriatic, especially in Croatia where 1,246 are found. The number includes islands, islets, and rocks of all sizes, including ones emerging at ebb tide only.[9] The Croatian islands include the largest—Cres and Krk each covering 405.78 square kilometres (156.67 square miles), and the tallest—Brač whose peak reaches 780 metres (2560 feet) above sea level. The Croatian islands include 48 permanently inhabited ones, the most populous among them being Krk and Korčula.[10] Islands found along the Western coast of the Adriatic Sea are smaller and less numerous in comparison to those along the opposite coast. The best known Adriatic islands found along the Western, i.e. Italian coast are 117 islands on which the city of Venice is built.[11] Northern shore of Greek island of Corfu also lies in the Adriatic sea, as defined by the IHO.[12] The IHO boundary also places few smaller Greek islands northwest of Corfu in the Adriatic Sea.[2] [13]

Adriatic Sea (Croatia)

Bathymetry

North Adriatic basin, extending between Venice and Trieste towards a line connecting Ancona and Zadar, is only 15 metres (49 feet) deep at its northwestern end, and it gradually deepens towards the southeast. Middle Adriatic basin extends south of the Ancona–Zadar line, as 270-metre (890 ft) deep Middle Adriatic Pit (also called Pomo Depression or Jabuka Pit). 170-metre (560 ft) deep Palagruža Sill is south of the Middle Adriatic Pit, separating it from 1200-metre (3900 ft) deep South Adriatic Pit and the Middle Adriatic basin from the South Adriatic Basin. Further on to the south, the sea floor rises to 780 metres (2560 feet) to form Otranto Sill at the boundary to the Ionian Sea. Transversally, the Adriatic Sea is also asymmetric as the Apennine peninsula coast is relatively smooth with very few islands and promontory of Monte Conero and Gargano Promontory as only significant protrusions into the sea. On the other hand, the Balkan peninsula coast is rugged with numerous islands, especially in Croatia. Ruggedness of the coast is exacerbated by proximity of the Dinaric Alps to the coast, in contrast to the opposite, Italian coast, where the Apennine Mountains are further away from the shoreline.[14] Average depth of the Adriatic sea is 252 metres (827 feet), while its maximum depth is 1233 metres (4045 feet). Still, the North Adriatic basin rarely exceeds depth of 100 metres (330 feet).[7]

Hydrology

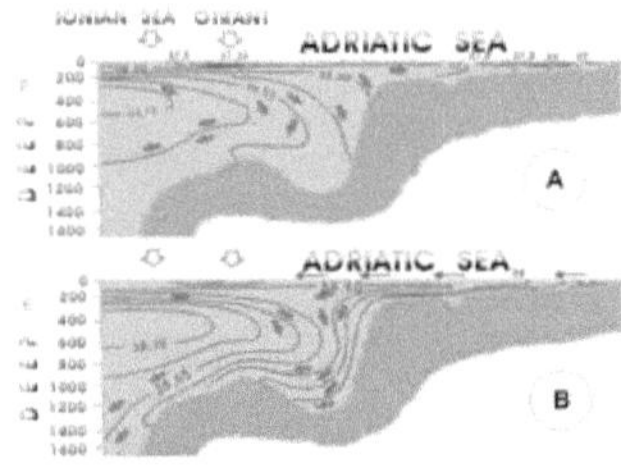

Currents and salinity on cross-sections of the East (A) and the West (B) Adriatic

Dynamics of coastal waters are determined by the asymmetric coasts and inflow of Mediterranean seawater through the Straits of Otranto and further on along the Eastern coast.[15] The smooth Italian coast, comprising very few protrusions into the sea and no major islands allows smooth flow of the Western Adriatic Current, while coastal currents on the opposite shore are far more complex because of jagged appearance of the shoreline, several large islands and proximity of the Dinaric Alps to the shore. The latter produces significant temperature variations between the sea and hinterland, which is conductive surrounding to creation of local jets.[14] Tidal movement is slight, normally remaining below 30 centimetres (12 inches). The amphidromic point is at mid-width east of Ancona.[16]

The normal tide levels are known to increase abnormally in conductive environment, leading to coastal flooding, most famously known in Italy, especially Venice, as acqua alta. They can exceed normal tide levels by more than 140 centimetres (55 inches), with the highest tide level of 194 centimetres (76 inches) observed on 4 November 1966. Such flooding is caused by a combination of factors, including alignment of the Sun and the Moon, geometric shape of the basin, which amplifies or reduces the astronomical component and meteorological factors such as atmospheric pressure and winds. Also, the long and narrow rectangular shape of the Adriatic Sea is the source of an oscillating water motion—seiche along the basin's minor axis[17] Finally, Venice is increasingly vulnerable to flooding because of subsidence of coastal area soil.[18] [19] Such floods were also observed elsewhere in the Adriatic Sea, and have been recorded in recent years in Koper, Zadar and Šibenik as well.[20] [21] [22]

It is estimated that the entire volume of the Adriatic Sea is exchanged through the Strait of Otranto in 3.4±0.4 years, which is a comparably short period. For instance, approximatey 500 years are necessary to exchange all water of the Black Sea. This is particularly important as rivers flowing into the Adriatic discharge up to 5700 cubic metres per second (cubic feet per second). That rate of discharge amounts to 0.5% of total Adriatic Sea volume or a 1.3-metre (4 ft 3 in) layer of water each year. The greatest portion of the discharge from any single river comes from the Po (28%),[23] whose average discharge alone is 1569 cubic metres per second (55400 cubic feet per second).[24] In terms of annual discharge in the Mediterranean Sea, Po is ranked the second, followed by Neretva and Drin which are rank as the third and

A submarine spring near Omiš, observed through rippling of the sea surface

the fourth.[25] Another significant contributor of freshwater to the Adriatic is submarine groundwater discharge (SGD) through submarine springs (Croatian: *vrulja*). The SGD is estimated to comprise 29% of total water flux in the Adriatic.[26] The submarine springs also include thermal springs, discovered off shore near the town of Izola. The thermal water is rich with hydrogen sulfide, has the temperature of 22 to 29.6 °C (72 to 85 °F), and has enabled the development of specific ecosystems.[27] The inflow of freshwater, representing one third of freshwater volume flowing into the Mediterranean,[28] gives the Adriatic characteristics of a dilution basin of the Mediterranean Sea.[29]

Temperature and salinity

Surface temperature of the Adriatic sea largely ranges from 22 to 24 °C (72 to 75 °F) in summer, or 12 to 14 °C (54 to 57 °F) in winter, except along the northern part of the western Adriatic coast, where the surface temperature drop to 9 °C (48 °F). The seasonal temperature variations are attributed to heat flux exchanged with the atmosphere. Salinity of the Adriatic sea ranges between 38 and 39 PSUs.[30] In shallow coastal areas of the Adriatic sea ice may appear in particularly cold winters—especially in the Venetian Lagoon,[31] but also in isolated shallow areas as far south as Tisno south of Zadar.[32]

Climate

As in most of the areas in the Mediterranean Basin, the Adriatic Sea and its surrounding landmass enjoy Mediterranean climate, a particular variety of subtropical climate. Since the Adriatic Sea is located at mid-latitudes, it is characterized by seasonal variability of climate. The climate is characterized by warm to hot, dry summers and mild to cool, wet winters. Air temperature fluctuates by about 20 °C (68 °F) during a season.[30] According to the Köppen climate classification, the southern and central Adriatic are classified as hot-summer Mediterranean climate (Csa), and the northern Adriatic as Humid subtropical climate (Cfa) climate areas.[33] [34] Predominant winter winds are bora and sirocco (*jugo*). Bora is significantly conditioned by wind gaps in the Dinaric Alps bringing cold and dry continental air, and it reaches peak speed in areas of Trieste, Senj, and Split, with gusts of up to 180 km/h (100 kn; 110 mph). Sirocco, conversely, brings humid and warm air often carrying Saharan sand and causing rain dust.[35]

Climate characteristics in major Adriatic cities									
City	Mean temperature (daily high)				Mean total rainfall				
	January		July		January			July		
	°C	°F	°C	°F	mm	in	days	mm	in	days
Bari	12.1	53.8	28.4	83.1	50.8	2.00	7.3	27.0	1.06	2.6
Dubrovnik	12.2	54.0	28.3	82.9	95.2	3.75	11.2	24.1	0.95	4.4
Rijeka	8.7	47.7	27.7	81.9	134.9	5.31	11.0	82.0	3.23	9.1
Split	10.2	50.4	29.8	85.6	77.9	3.07	11.1	27.6	1.09	5.6
Venice	5.8	42.4	27.5	81.5	58.1	2.29	6.7	63.1	2.48	5.7
Source:World Meteorological Organization[36]									

Population

Most populous urban areas on the Adriatic Sea coast

	Rank	City	Country	Region/County	Population (urban)	
Bari **Venice**	1	**Bari**	Italy	Apulia	320,475	**Trieste** **Split**
	2	**Venice**	Italy	Veneto	270,884	
	3	**Trieste**	Italy	Friuli-Venezia Giulia	205,535	
	4	**Split**	Croatia	Split-Dalmatia	165,893	
	5	**Ravenna**	Italy	Emilia-Romagna	159,497	
	6	**Rimini**	Italy	Emilia-Romagna	142,579	
	7	**Rijeka**	Croatia	Primorje-Gorski Kotar	127,498	
	8	**Pescara**	Italy	Abruzzo	123,103	
	9	**Durrës**	Albania	Durrës	115,550	
	10	**Ancona**	Italy	Marche	101,210	
	Sources: 2011 Croatian census[37] , Italian National Institute of Statistics (2011)[38] , Albanian Census 2011[39]					

On the coasts and islands of the Adriatic Sea there are numerous settlements, but there are few larger cities. Among the largest there are Bari, Venice, Trieste, Ravenna and Rimini in Italy, Split, Rijeka and Zadar in Croatia, Durrës and Vlorë in Albania and Koper in Slovenia. In total, more than 3.5 million people live at the Adriatic coasts.[40]

Coastal management

Venice, which was originally built on islands off the coast, is most at risk due to subsidence, but the effect is realized in the Po delta as well. The causes are a decrease in the sedimentation rate due to loss of sediment behind dams and the deliberate excavation of sand for industrial purposes agricultural use of water, and removal of ground water.[41] [42]

Sinking of the city slowed since use of the artesian wells was banned in the 1960s, but the city remains threatened by the acqua alta floods. Recent studies suggested that the city is no longer sinking,[43] [44] but a state of alert remains in place. In May 2003 the Italian Prime Minister

MOSE Project north of Lido di Venezia

Silvio Berlusconi inaugurated the MOSE project (Italian: *Modulo Sperimentale Elettromeccanico*), an experimental model for evaluating the performance of inflatable gates. The project proposes laying a series of 79 inflatable pontoons across the sea bed at the three entrances to the Venetian Lagoon. When tides are predicted to rise above 110 centimetres (43 inches), the pontoons will be filled with air and block the incoming water from the Adriatic Sea. This engineering work is due to be completed by 2014.[45]

Geology

Adriatic Microplate boundaries

Geophysical and geological information indicate that the the Adriatic Sea and the Po Valley are are associated with a microplate which separated from the African Plate—identified as Apulian or Adriatic Plate during the Mesozoic era. The separation began in Middle and Late Triassic, when limestone begins sedimenting in the area. Between Norian and Late Cretaceous, Adriatic and Apulia Carbonate Platforms form as thick series of carbonate sediments (dolomites and limestones), up to 8000 metres (26000 feet) deep.[46] Remnants of the former are found in the Adriatic Sea itself, as well as in the southern Alps and the Dinaric Alps, and remnants of the latter are exhibited as the Gargano Promontory and the Maiella Mountain. In the Eocene and early Oligocene the plate moves to the North and Northeast, contributing to Alpine orogeny, specifically orogeny of the Dinarides and the Alps. In Late Oligocene, the motion is reversed and the Apennine Mountains orogeny took place.[47] An unbroken zone of increased seismic activity borders the Adriatic Sea, with a belt of thrust faults generally oriented Northeast–Southwest on the East coast and Northeast–Southwest normal faults in the Apennines—indicating counterclockwise rotation of the Adriatic.[48] An active 200-kilometre (120 mi) fault was identified Northwest of Dubrovnik, adding to the Dalmatian islands as the Eurasian Plate slides over the Adriatic microplate. Furthermore, the fault causes the southern tip of the Apennine peninsula to move towards the opposite shore by about 0.4 centimetres (0.16 inches) per year. If that movement continues, the seafloor will be completely consumed and the Adriatic Sea closed off in 50–70 million years.[49] In the Northern Adriatic, the coast of the Gulf of Trieste and western Istria is gradually subsiding, sinking about 1.5 metres (4 feet 11 inches) in past two thousand years.[50]

Seafloor sediment

All types of seafloor sediments are found in the Adriatic Sea. Comparably shallow seabed of the Northern Adriatic is characterized by relict sand, while muddy bed is typical at depths below 100 metres (330 feet).[15] There are five geomorphological units in the Adriatic: Northern Adriatic (up to 100 metres (330 feet) deep), North Adriatic islands area protected against sediments filling it in by outer islands (pre-Holocene karst relief), Middle Adriatic islands area (large Dalmatian islands), Middle Adriatic (characterized by the Middle Adriatic Depression) and the Southern Adriatic consisting of a coastal shelf and the Southern Adriatic Depression. Sediments deposited in the Adriatic Sea today generally originate at the Northwest coast, carried by rivers Po, Reno, Adige, Brenta, Tagliamento, Piave and Soča. Volume of sediments carried from the Eastern shore by rivers Rječina, Zrmanja, Krka, Cetina, Ombla, Dragonja, Mirna, Raša and Neretva is negligible as those are largely deposited at the river mouths.

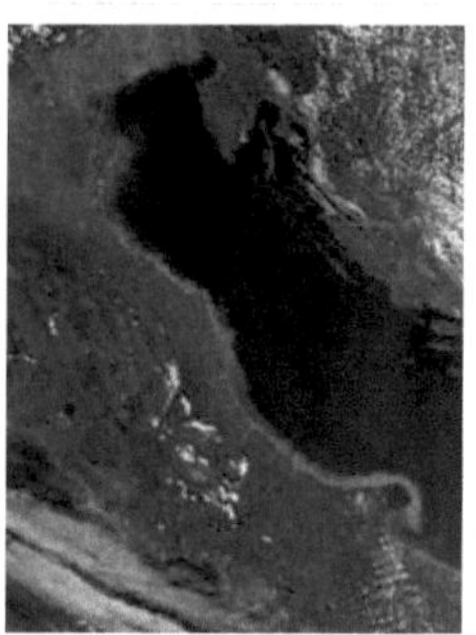

Sediment billowing out from Italy's shore into the Adriatic.

The western shores of the Adriatic are largely either alluvial or terraced, while the eastern shores are predominantly rocky, except for the southernmost part of the shore located in Albania, which consists of sandy coves and rocky capes.[47]

Coasts

The Eastern Adriatic coast, especially between Gulf of Trieste and Bay of Kotor, is one of the most indented ones. The majority of the eastern coast is characterized by karst topography, developed from the Adriatic Carbonate Platform. Karstification there largely began after final uplift of the Dinarides in Oligocene and Miocene, when the carbonate deposits were exposed to atmospheric effects, extending to level of 120 metres (390 feet) below present sea level, exposed during the Last Glacial Maximum. It is estimated that some karst formations are related to earlier emersions, most notably the Messinian salinity crisis.[46] Similarly, karst developed in Apulia from the Apulian Carbonate Platform.[51]

The largest part of the Eastern coast consists of carbonate rocks, while flysch is significantly represented in the Gulf of Trieste coast, especially along the Slovenia's coast where 80-metre (260 ft) Strunjan cliff, the highest on the entire Adriatic coast is located,[52] on the Kvarner Gulf coast opposite Krk, and in Dalmatia north of Split.[53] The same type of rocks are found in Albania,[54] and on the Western Adriatic coast.[55]

There are alternations of maritime and alluvial sediments occurring in Po Valley, at the Northwest coast of the Adriatic, as far west as Piacenza, dating to Pleistocene as the sea advanced and receded over the valley. An advance began after the Last Glacial Maximum, which brought the Adriatic to a high point at about 5,500 years ago.[56] Since then the Po delta had been prograding. The rate of coastal zone progradation between 1000 BC and 1200 AD was 4 metres (13 feet) per year.[57] In the 12th century, the delta advanced at a rate of 25 metres (82 feet) per year. In 1600s, the delta began becoming a human-controlled environment, as excavation of artificial channels started. The channels and new distributaries of Po prograded at rates of 50 metres (160 feet) per year or more since.[58] There are more than 20 other rivers flowing into the Adriatic Sea also forming alluvial coastlines in Italy alone.[59] There are comparably smaller alluvial coasts of the Eastern Adriatic—in deltas of Dragonja,[60] Bojana,[61] and Neretva.[62]

Flora and fauna

The Adriatic Sea is unique unit of the Mediterranean in terms of biogeography, particularly because of numerous endemic species. Croatian National Biodiversity Strategy Action Plan identified more than 7,000 animal and plant species in the Adriatic Sea. The Middle Adriatic basin is especially abundant in endemic plant species, with 535 identified species of green, brown and red algae.[63] Four out of five Mediterranean seagrass species are found in the Adriatic Sea. The most common species are Cymodocea nodosa and Zostera noltii, while Zostera marina and Posidonia oceanica are comparably rare.[64]

A number of rare and threatened species are also found along the eastern coast of the Adriatic, and as it is relatively clearer and less polluted compared to the Western Adriatic coast—in part because the sea currents flow through the Adriatic in counterclockwise direction, thus bringing clearer waters up the Eastern coast and returning as increasingly polluted water down the Western coast. This has significantly contributed to the biodiversity of the countries along the eastern Adriatic coast. The common bottlenose dolphin is common in the Eastern coast waters only. Approximately thirty species of fish are found only in one or two of the countries bordering Adriatic Sea. Those are particularly related to the karst morphology of the coastal or submarine topography. There are 45 subspecies endemic to the Adriatic Sea coast and islands. In the Adriatic Sea, there are at least 410 species and subspecies of fish, representing approximately 70% of Mediterranean taxa, with seven species endemic to the Adriatic. 64 species are considered threatened, largely because of overfishing.[63] Only a small fraction of the fish fauna found in the Adriatic is attributed to recent processes such as Lessepsian migration, cases of escape from mariculture and similar.[65]

Protected areas

Isole Tremiti protected area

Even though marine biodiversity of the Adriatic Sea is relatively high, several marine protected areas were established by some of the countries along its coasts. In Italy, those are Miramare in the Gulf of Trieste, in the Northern Adriatic, Torre del Cerrano and Isole Tremiti in the Middle Adriatic basin and Torre Guaceto in southern Apulia.[66] The Miramare protected area was established in 1986 and covers 30 hectares (74 acres) of coast and 90 hectares (220 acres) of sea. The area encompasses 1.8 kilometres (1.1 miles) of coastline near Miramare promontory in the Gulf of Trieste.[67] The Torre Cerrano protected area was created in 2009, extending 3 nautical miles (5.6 kilometres) into the sea and along 7 kilometres (4.3 miles) of coastline. Various zones of the protected area cover 37 square kilometres (14 square miles) of sea surface only.[68] Isole Tremiti are a protected area since 1989, while the Tremiti islands themselves are a part of the Gargano National Park.[69] The Torre Guaceto protected area, located near Brindisi and Carovigno covers sea surface of 2227 hectares (5500 acres) and it is adjacent to Torre Guaceto State Reserve covering 1114 hectares (2750 acres) of coast and sharing a 8-kilometre (5.0 mi) coastline with the marine protected area.[70] Furthermore there are 10 Ramsar wetland reserves in Italy located along the Adriatic coast.[71]

There are seven marine protected areas in Croatia: Brijuni and Lim Canal off coast of Istria peninsula, near Pula and Rovinj respectively, Kornati and Telašćica in the Middle Adriatic basin, near Šibenik, and Lastovo, Bay of Mali Ston (Croatian: *Malostonski zaljev*) and Mljet in the Southern Dalmatia.[66] Brijuni national park encompasses the 743.3-hectare (1837-acre) archipelago itself and 2651.7 hectares (6552 acres) of surrounding sea.[72] Brijuni became a national park in 1999.[73] Lim Canal, a 10-kilometre (6.2 mi) ria of Pazinčica river.[74] Kornati national park was established in 1980 and it covers approximately 220 square kilometres (85 square miles), including 89 islands and islets. Still, the marine environment encompasses three

Kornati national park

quarters of the total area, while length of the island shores combined equals 238 kilometres (148 miles).[75] Telašćica is a nature park established on Dugi Otok in 1988. The park covers 69-kilometre (43 mi) coastline, 22.95 square kilometres (8.86 square miles) of land and 44.55 square kilometres (17.20 square miles) of sea.[76] Bay of Mali Ston is located at the border of Croatia and Bosnia and Herzegovina, north of Pelješac peninsula. The marine protected are covers 48 square kilometres (19 square miles).[66] Lastovo nature park was established in 2006, and it includes 44 islands and islets, 53 square kilometres (20 square miles) of land and 143 square metres (1540 square feet) of sea surface.[77] Mljet national park was established in 1960, covering 24-square-kilometre (9.3 sq mi) marine protection area.[66] In addition, there is a Ramsar wetland reserve in Croatia—Neretva River delta.[78]

In Slovenia, the marine and coastal protected nature areas are the Sečovlje Salina Nature Park, the Strunjan Landscape Park, the Škocjan Inlet Nature Reserve, and the Debeli Rtič, the Cape Madona and Lakes in Fiesa natural monuments.[79] [80] The Sečovlje Salina Nature Park was established in 1990, covers 721 hectares (1780 acres), and includes four nature reserves.[81] [82] In 1993, the area has been designated a Ramsar site,[79] and is also a site of international importance for waterbird species.[83] The 429 ha (1060-acre) Strunjan Landscape Park was established in 2004,[81] comprises two nature reserves, and includes a 4 km (2.5 mi) long cliff, the northernmost Mediterranean salt field and the only Slovenian lagoon system.[84] The Škocjan Inlet Nature Reserve was established in 1998 and covers 122 hectares (300 acres).[85] The Debeli Rtič natural monument covers 25 hectares (62 acres),[86] the Cape Madona natural monument covers 13 hectares (32 acres),[86] and the coastal lake in Fiesa, the only brackish lake in

Slovenia, covers 2.5 hectares (6.2 acres).[87]

In 2010, Albania established its first marine protection area of the Karaburun-Sazan National Marine Park at the Karaburun Peninsula, where the Adriatic and the Ionian seas meet. The park covers a total of 12570 hectares (31100 acres).[88] Two additional marine protection areas are planned in Albania: Cape of Rodon (Albanian: *Kepi i Rodonit*) and Porto Palermo.[66] In addition, Albania is home to two Ramsar wetland reserves[89] Neither Bosnia and Herzegovina nor Montenegro plan to establish any marine protection areas.[66]

History

Name

Etymology of the Adriatic Sea is linked to Etruscan settlement of Adria, itself probably originating from Illyrian word *adur* meaning water or sea.[90] In classical antiquity, the sea was known as *Mare Adriaticum* (*Mare Hadriaticum*, also sometimes simplified to *Adria*) or, less frequently, as *Mare Superum*.[91] The two terms were not synonymous, however. *Mare Adriaticum* generally corresponds to extent of the Adriatic Sea, spanning from the Gulf of Venice to the Strait of Otranto. That boundary became more consistently defined by Roman authors—early Greek sources place boundary between the Adriatic and Ionian Seas at various places ranging from adjacent to the Gulf of Venice to the southern tip of Peloponnese, eastern shores of Sicily and western shores of Crete.[92] *Mare Superum* on the other hand normally encompassed modern Adriatic Sea and the sea off the southern coast of Apennine peninsula, as far as the Strait of Sicily.[93] Another name used in the period was *Mare Dalmaticum*, applied to waters off coast of Dalmatia or Illyricum.[94]

Early history

Settlements along the Adriatic appear in Albania and Dalmatia on the eastern coast dating to between 6100 and 5900 BC, related to the Cardium Pottery Culture.[95] During the Classical Antiquity the shores of the Adriatic were initially inhabited by the Ancient peoples of Italy—among whom the Etruscan civilization becoming the most prominent before rise of the Roman Republic—on the Western Adriatic coast,[96] and by the Illyrians along the Eastern Adriatic coast.[97] Greek colonisation of the Adriatic dates back to the 7th and the 6th century BC when Epidamnus and Apollonia were founded. Greeks soon expanded further north establishing several cities, including Epidaurus, Black Corcyra, Issa and Ancona, with trade established as far North as the Po River delta, where emporion of Adria

Pula Arena, one of the six largest surviving Roman amphitheatres

was founded. Following the Roman intervention and the Illyrian Wars, the Eastern Adriatic shore became a province of the Roman Republic.[98] The initial intervention in 229 BC marked the first time that the Roman navy crossed the Adriatic to launch a military campaign.[99] In Middle Ages, after the decline of the Roman Empire, coasts of the Adriatic were ruled by Ostrogoths, Lombards,[100] and the Byzantine Empire.[101] The last part of the Early Middle Ages saw rise of the Carolingian Empire and subsequent Frankish Kingdom of Italy which controlled the Western coast of the Adriatic Sea,[102] while the Byzantine control of the opposite coast gradually shrunk following Avar and Croatian invasions starting in the 7th century.[103] Republic of Venice was founded in the period, and it later became a significant trading power after receiving Byzantine tax exemption in 1082.[104] The end of the period brought about Holy Roman Empire's control over the Kingdom of Italy which would last until the Peace of Westphalia in 1648,[105] establishment of an independent Kingdom of Croatia,[106] and Byzantine Empire expanded to the Southern Apennine peninsula.[107] Furthermore, the Papal States were carved out in areas around Rome in the 8th century.[108]

Venice was a leading trading power in Europe

High Middle Ages in the Adriatic Sea basin saw further territorial changes, including the Norman conquest of southern Italy ending Byzantine presence on the Apennine peninsula in the 11th and the 12th centuries,[109] , and the territory would become the Kingdom of Naples in 1282,[110] and control of a substantial part of the Eastern Adriatic coast by the Kingdom of Hungary after a personal union was established between Croatia and Hungary in 1102.[111] In the period, the Republic of Venice began to expand its territory and influence.[107]

In 1202, the Fourth Crusade was diverted to conquer Zadar on behest of Venetians—the first instance of a Crusader force attacking a Catholic city—before proceeding to sack Constantinople.[112] In the century, Venice established itself as one of the leading trading nations. During much of the 12th and the 13th century, Venice and the Republic of Genoa were engaged in Venetian–Genoese Wars ending in War and Battle of Chioggia, removing Geonoese from the Adriatic.[113] Still, the Treaty of Turin of 1381 that ended the war required Venice to renounce claims to Dalmatia, after losing the territory to Hungary in 1358. In the same year, the Republic of Dubrovnik was established a city-state as it escaped Venetian suzerainty.[114] Venice regained Dalmatia in 1409 and held it for nearly four hundred years—reaching the republic's apex of power in the first half of the 15th century.[115] The 15th and the 16th centuries brought about destruction of the Byzantine Empire (1453),[116] expansion of the Ottoman Empire which reached Adriatic shores in present-day Albania and Montenegro,[117] as well as immediate hinterland of Dalmatian coast, defeating the Hungarian and Croatian armies at Krbava (1493) and Mohács (1526).[118] Those defeats spelled the end of an independent Hungarian kingdom, and both Croatian and Hungarian nobility chose Ferdinand I of the House of Habsburg as the new ruler and bringing the Habsburg Monarchy to the shore of Adriatic Sea, where it will remain for nearly four hundred years.[119] The Ottomans and the Venetians fought a series of wars, but until the 17th century, those were not fought in the Adriatic area.[115] The Ottoman raids of the Adriatic coasts were effectively stopped after the Battle of Lepanto (1571) fought south of the Strait of Otranto.[120]

Age of sail

In 1648, the Holy Roman Empire lost its claim on the Italian lands, formally ending the Kingdom of Italy, however its only outlet on the Adriatic Sea, the Duchy of Ferrara was already lost to the Papal States.[121] The final territorial changes of the 17th century were caused by the Morean or the Sixth Ottoman-Venetian War, when Venice slightly enlarged its possessions in Dalmatia in 1699.[122] In 1797, the Republic of Venice was abolished after French conquest (1797).[123] The Venetian territory was then handed over to Austria and briefly ruled as a part of Archduchy of Austria. The territory was turned back to France after the Peace of Pressburg (1805) when the territory in the Po Valley became an integral part of new Kingdom of

Battle of Lissa, 1811

Italy.[124] The Kingdom included Romagna province thus removing the Papal State from the Adriatic coast,[125] but Trieste, Istria and Dalmatia were made a separate province of the French Empire—the Illyrian Provinces.[124] The Illyrian Provinces were created in 1809 through the Treaty of Schönbrunn and they represented the end of Venetian rule on the East Adriatic coast, as well as the end of existence of the Republic of Dubrovnik.[126] The Adriatic Sea was a minor theatre of war during the Napoleonic Wars, as the Adriatic campaign of 1807–1814 involved British Royal Navy contesting control of the Adriatic by the combined navies of France, Italy and Kingdom of Naples. During the campaign, the Royal Navy occupied Vis and established its base there in Port St. George.[127] The campaign reached its climax in the Battle of Lissa in 1811,[128] and ended in British troops and Austrian army seizing the Eastern Adriatic coast cities from the French.[129] Days before the battle of Waterloo, the Congress of Vienna awarded the Illyrian Provinces spanning from the Gulf of Trieste to the Bay of Kotor to Austria.[130] The Congress of Vienna also created the Kingdom of Lombardy–Venetia, encompassing the city of Venice, surrounding coast and substantial hinterland—ruled by Austria.[131] In the south of the Apennine peninsula, the Kingdom of the Two Sicilies was formed in 1816 through a union of the Kingdom of Naples and the Kingdom of Sicily.[132]

From ironclads to dreadnoughts

Battle of Lissa, 1866

Process of Italian unification culminated in the Second Italian War of Independence resulting in Kingdom of Sardinia annexing all territories along the Western Adriatic coast south of Venetia in 1860, and establishment of the Kingdom of Italy in 1861 in its place. The Kingdom of Italy expanded in 1866. It annexed Venetia,[133] but Regia Marina was defeated in the Adriatic near Vis.[134] Following the Austro-Hungarian Compromise of 1867 and Croatian–Hungarian Settlement of 1868, control of much of the Eastern Adriatic coast was redefined. Cisleithanian part of the Austria-Hungary spanned from Austrian Littoral to the Bay of Kotor, with exception of the Croatian Littoral. A Corpus separatum formed in 1779 and containing the city of Rijeka directly subjected to the Kingdom of Hungary was confirmed and the rest was a part of the Kingdom of Croatia-Slavonia, which in turn was also in the Transleithanian part of the dual monarchy.[111] Length of the Adriatic coastline controlled by the Ottoman Empire shrunk in 1878 when the Congress of Berlin recognized independence of the Principality of Montenegro controlling the coast south of Bay of Kotor to Bojana River.[135] The Ottoman Empire was completely removed from the Adriatic following the First Balkan War and consequent Treaty of London of 1913 which established independent Albania.[136]

World War I Adriatic Campaign was largely limited to attempts of blockade of the sea by the Allies and Central Powers' attempts to break it.[137] Italy joined the Allies in 1915 after the Treaty of London promised it acquisition of the Austrian Littoral, Northern Dalmatia, port of Vlorë, most East Adriatic islands and establishment of a protectorate in Albania.[138] In 1918, Montenegrin national assembly voted to form an union with the Kingdom of Serbia, giving the latter access to the Adriatic.[139] Another short-lived, unrecognized state established in 1918 was the State of Slovenes, Croats and Serbs out of

SMS *Novara* after Battle of the Strait of Otranto, 1917

parts of Austria-Hungary, comprising most of the coastline of the former monarchy. Later that year, the two formed the Kingdom of Serbs, Croats and Slovenes—subsequently renamed to Yugoslavia. Proponents of the new union in the Parliament of Croatia at the time saw the move as a defence against Italian expansionism and provisions of the Treaty of London.[140] The treaty was largely disregarded by the Britain and France because of conflicting promises made to Serbia and perceived lack of Italian contribution to the war effort outside Italy itself.[141] The Treaty of Saint-Germain-en-Laye of 1919 did transfer the Austrian Littoral and Istria to Italy, but awarded Dalmatia to Yugoslavia.[142] Following the war, a private force of demobilized Italian soldiers seized Rijeka and set up Italian Regency of Carnaro, seen as harbinger of Fascism, in order to force recognition of Italian claim to the city.[143] After sixteen months of its existence, the Treaty of Rapallo of 1920 redefined Italian–Yugoslav borders, inter alia transferring Zadar and islands of Cres, Lastovo and Palagruža to Italy, securing the island of Krk for Yugoslavia and establishing the Free State of Fiume. The Free State of Fiume was abolished in 1924 by the Treaty of Rome which awarded Rijeka to Italy and Sušak to Yugoslavia.[144]

Modern era

Boundary between Free Territory of Trieste and Italy

In the World War II, the Adriatic saw limited naval action starting with Italian invasion of Albania and Axis Invasion of Yugoslavia. The latter resulted in annexation of a large part of Dalmatia and nearly all East Adriatic islands to Italy and establishment of puppet states of Independent State of Croatia and Kingdom of Montenegro controlling the remainder of former Yugoslav Adriatic coast.[145] In 1947, after the Armistice between Italy and Allied armed forces and the end of the war, Italy, now a republic, and Allies of World War II signed Treaty of Peace with Italy reversing all wartime annexations, transferring islands of Cres, Lastovo and Palagruža, the cities of Zadar and Rijeka, Istria and most of Slovenian Littoral to communist Yugoslavia, guaranteeing independence of Albania and carving out Free Territory of Trieste (FTT) as a city-state.[146] The FTT was partitioned in 1954 as Trieste itself and area to the North of it were placed under Italian and the rest under Yugoslav control. The arrangement was made permanent by the Treaty of Osimo in 1975.[147]

During the Cold War, the Adriatic Sea became the southern flank of the Iron Curtain as Italy allied itself with NATO,[148] while the Warsaw Pact established bases in Albania.[149] After the fall of communism, Yugoslavia broke apart as Slovenia and Croatia declared independence in 1991,[150] and Bosnia and Herzegovina followed in 1992,[151] while Montenegro remained in a federation with Serbia, subsequently renamed to Serbia and Montenegro.[152] The Croatian War of Independence that ensued included limited naval engagement and maritime blockade of Croatian coast by the Yugoslav Navy,[153] resulting in the Battle of the Dalmatian channels and withdrawal of Yugoslav vessels.[154] Montenegro declared its independence in 2006, ending Yugoslav control of any portion of the Adriatic coast.[152] The period also saw Adriatic Sea as the theatre of several NATO operations, including blockade of Yugoslavia,[155] , intervention in Bosnia and Herzegovina,[156] and bombing of Yugoslavia.[157]

Boundaries

Italy and Yugoslavia defined their delimitation of the continental shelf in the Adriatic Sea in 1968,[158] with an additional agreement on the boundary in the Gulf of Trieste signed in 1975 pursuant to the Treaty of Osimo. The boundary agreed in 1968 extends 353 nautical miles (654 kilometres; 406 miles) and consists of 43 points connected by straight lines or circular arc segments. The boundary agreed upon in 1975 consists of 5 points, extending from an end point of the 1968 line. All successor states of the former Yugoslavia accepted the agreements. In the southernmost areas of the Adriatic the border was not determined in order to avoid prejudicing location of tripoint with the Albanian continental shelf border, which remains undefined. Prior to

Signing ceremony of the Treaty of Osimo, 10 November 1975

breakup of Yugoslavia, Albania, Italy and Yugoslavia initially proclaimed 15-nautical-mile (28 km; 17 mi) territorial waters, subsequently reduced to 12 nautical miles (22 kilometres; 14 miles) and all sides adopted baseline systems. Most of these were introduced in 1970s. Albania and Italy determined their border at the sea in 1992 on equidistance principle.[159] Following accession of Croatia to the EU, the Adriatic is expected to become an internal sea of the EU.[160]

Adriatic Euroregion

Adriatic Euroregion was established in Pula in 2006 to promote trans-regional and trans-national cooperation in the Adriatic Sea area, representing an Adriatic framework to help resolve issues of regional importance. The Adriatic Euroregion consists of 23 members—Apulia, Molise, Abruzzo, Marche, Emilia-Romagna, Veneto and Friuli-Venezia Giulia regions of Italy, municipality of Izola in Slovenia, Istria, Primorje-Gorski Kotar, Lika-Senj, Zadar, Šibenik-Knin, Split-Dalmatia and Dubrovnik-Neretva counties of Croatia, Herzegovina-Neretva Canton of Bosnia and Herzegovina, municipalities of Kotor and Tivat in Montenegro, Fier, Vlorë, Tirana, Shkodër, Durrës and Lezhë counties of Albania and Greek prefectures of Thesprotia and Corfu.[161]

Disputes

Unlike land borders of former Yugoslav republics and subsequently the successor states, their maritime borders were not defined. In addition, the maritime boundary between Albania and Montenegro was not defined before the 1990s. Croatia and Slovenia started negotiations in 1992 but failed to agree resulting in a dispute which was referred to arbitration in 2009. Croatia also declared Ecological and Fisheries Protection Zone (ZERP) extending to the continental shelf boundary, but its application to the EU member states was suspended pending definition of the borders. In 2005, Slovenia also declared its protected ecological and continental shelf zone encompassing parts of the Croatia's ZERP. Still the dispute causes no major practical problems in reality. Maritime boundary between Bosnia and Herzegovina and Croatia was formally settled in 1999, but few issues remained in dispute—Klek peninsula and two islets in the border area. Croatia–Montenegro maritime boundary is disputed in the Bay of Kotor, at the Prevlaka peninsula, exacerbated by occupation of the peninsula by the Yugoslav People's Army and later by FR Yugoslav Army, ultimately replaced by an United Nations observer mission. The observers left in 2002 and Croatia took over the area as an agreement allowing Montenegrin presence in Croatian waters in the bay was made. The dispute proved far less contentious since independence of Montenegro.[159]

Economy

Fishing

The Adriatic Sea fisheries production varies among countries in the basin. In 2000, nominal total landings of all Adriatic fisheries reached 110 thoushand tonnes.[162] The largest volume of the fisheries production is achieved in Italy, where the total production volume in 2003 stood at 472 thousand tonnes for human consumption and 15 thousand tonnes for other purposes. 28.8% of that volume are landings in the Northern and central Adriatic, and 24.5% in Apulia—Southern Adriatic and Ionian Sea. Italian fisheries, including those operating in outside the Adriatic, employ 60,700 in the primary sector, including aquaculture which comprises 40% of the total fisheries production. Gross value of the total fisheries output in 2002 was 1.9 billion US$.[163]

In 2006, total Croatian fisheries production volume was 37.8 thousand tonnes of catch and 14.2 thousand tonnes in marine aquaculture. Croatian fisheries, employ approximately 20,000. Marine capture catch in 2006 in Croatian waters consisted of sardines (44.8%), anchovies (31.3%), tunas (2.7%), other pelagic fish (4.8%), hake (2.4%), mullet (2.1%), other demersal fish (8.3%), crustaceans (largely lobster and Nephrops norvegicus) (0.8%), shellfish (largely oysters and mussels) (0.3%), cuttlefish (0.6%), squids (0.2%) and octopuses and other cephalopods (1.6%). Croatian marine aquaculture production consists of tuna (47.2%), oysters and mussels (28.2% combined) and bass and bream (24.6% combined).[164]

In 2006, Albanian fisheries production amounted to 7,699 tonnes, including 1,970 tonnes of aquaculture production. At the same time, Slovenian fisheries produced a total of 2,500 tonnes with 55% of the production volume originating in aquaculture, representing the highest ratio in the Adriatic. Finally, Montenegrin fisheries production stood at 911 tonnes in 2006, with all but 11 tonnes coming from capture production.[165]

Tourism

Countries bordering the Adriatic Sea are significant tourist destinations. The largest number of tourist overnight stays and the most numerous tourist accommodation facilities are recorded in Italy, especially in Veneto region. Veneto is followed by Emilia-Romagna region and by Adriatic Croatia counties. The Croatian tourist facilities are further augmented by 21 thousand nautical ports and moorings, as nautical tourists are attracted to various types of protected areas.[66] All countries along the Adriatic Sea coasts, except Albania and Bosnia and Herzegovina take part in the Blue Flag beach certification programme of the Foundation for Environmental Education for beaches and marinas meeting strict quality standards including environmental protection, water quality, safety and services criteria.[166] As of January 2012, the Blue Flag is awarded to 103 Italian Adriatic beaches and 29 marinas, 116 Croatian beaches and 19 marinas, 7 Slovenian beaches and 2 marinas, and 16 Montenegrin beaches.[167]

Tourism is also a significant source of income for those countries, especially so in Croatia and Montenegro where the tourist income generated along the Adriatic coast represents the bulk of tourism income.[168] [169] Direct contribution of travel and tourism to Croatian GDP stands at 5.1% in 2011, with total contribution of the industry estimated at 12.8% of the national GDP,[170] and in case of Montenegro, the direct contribution of tourism to national GDP is 8.1%, with the total contribution to economy at 17.2% of Montenegrin GDP.[171] Tourism in Adriatic Croatia is exhibiting greater growth recently, compared to the other regions around the Adriatic Sea.[172]

Dubrovnik is a major tourist destination in Croatia

Rimini is a major seaside tourist resort in Italy

Portorož is the largest seaside tourist centre in Slovenia

Tourism in the Adriatic Sea area[173] [174] [175] [168] [176] [177] [178]				
Country	Region	CAF beds*	Hotel beds	Overnight Stays
Albania	N/A	?	?	2,302,899
Bosnia and Herzegovina	Neum municipality	c. 6,000	1,810	280,000
Croatia	Adriatic Croatia	411,722	137,561	34,915,552

Italy	Friuli-Venezia Giulia	152,847	40,921	8,656,077
	Veneto	692,987	209,700	60,820,308
	Emilia-Romagna	440,999	298,332	37,477,880
	Marche	193,965	66,921	10,728,507
	Abruzzo	108,747	50,987	33,716,112
	Molise	11,711	6,383	7,306,951
	Apulia**	238,972	90,618	12,982,987
Montenegro	N/A	40,427	25,916	7,964,893
Slovenia	Seaside municipalities	24,080	9,330	1,981,141

*Beds in collective accommodation facilities, include "Hotel beds" figure also shown separately
**Includes both Adriatic and Ionian sea coasts

Transport

There are 19 major Adriatic Sea ports in four different countries, each handling more than a million tonnes of cargo per year. The largest cargo ports among them are the Port of Trieste—the largest Adriatic cargo port in Italy, the Port of Venice, the Port of Ravenna, Port of Koper—the largest Slovenian port,[179] Port of Rijeka—the largest Croatian cargo port, and the Port of Brindisi. The largest passenger ports in the Adriatic are the Port of Split;mdash;the largest Croatian passenger port, and ports in Ancona—the largest Italian passenger seaport in the Adriatic, Bari and Venice.[180] [181] [182] [183] The largest seaport in Montenegro is the Port of Bar.[184] In 2010, Norhtern Adriatic seaports of Trieste, Venice, Ravenna, Koper and Rijeka founded the North Adriatic Ports Association in order to position themselves more favourably in transport systems of the EU.[185] [186]

Port of Trieste, the largest cargo port in the Adriatic

Port of Rijeka, the largest cargo port in Croatia

Port of Durrës

Major Adriatic ports*, annual transport volume			
Port	**Country, Region/County**	**Cargo (tonnes)**	**Passengers**
Ancona	Italy, Marche	5,167,000	1,483,000
Bari	Italy, Apulia	3,197,000	1,392,000
Barletta	Italy, Apulia	1,390,000	N/A
Brindisi	Italy, Apulia	10,708,000	469,000
Chioggia	Italy, Veneto	2,990,000	N/A
Durrës	Albania, Durrës	3,441,000	770,000
Falconara Marittima	Italy, Marche	5,406,000	N/A
Koper	Slovenia, Slovenian Istria	17,100,000	100,300
Manfredonia	Italy, Apulia	1,277,000	N/A
Monfalcone	Italy, Friuli-Venezia Giulia	4,544,000	N/A
Ortona	Italy, Abruzzo	1,340,000	N/A
Ploče	Croatia, Dubrovnik-Neretva	5,104,000	146,000
Porto Nogaro	Italy, Friuli-Venezia Giulia	1,475,000	N/A
Rabac	Croatia, Istria	1,090,000	669,000
Ravenna	Italy, Emilia-Romagna	27,008,000	N/A
Rijeka	Croatia, Primorje-Gorski Kotar	15,441,000	219,800
Split	Croatia, Split-Dalmatia	2,745,000	3,979,439
Trieste	Italy, Friuli-Venezia Giulia	39,833,000	N/A
Venice	Italy, Veneto	32,042,000	1,097,000

*Ports handling more than a million tonnes of cargo or serving more than a million passengers per year
Sources: National Institute of Statistics (2010 data, Italian ports, passenger traffic below 200,000 is not reported),[180]
Croatian Bureau of Statistics (2008 data, Croatian ports, note: Port of Rijeka includes Rijeka, Bakar, Bršica and Omišalj terminals;
Port of Ploče includes Ploče and Metković terminals),[181]
Durrës' Chamber of Commerce and Industry - Albania (2007 data, Port of Durrës),[182] limun.hr (2011 data, Port of Koper)[183]

Oil and gas

Natural gas is produced through several projects, including a joint venture started by Eni and INA companies which operates two platforms—one of them is in Croatian waters and draws gas from six wells, and the other one, which started operating in 2010, is located in Italian waters. The Adriatic gas fields were discovered in the 1970s, but their development commenced in 1996. In 2008, INA produced 14.58 million BOE per day of gas.[187] Approximately a hundred offshore platforms are located in Emilia-Romagna region.[66] Eni estimated its concessions in the Adriatic Sea to hold at least 40000000000 cubic metres (1.4×10^{12} cubic feet) of natural gas, adding that they may even reach 100000000000 cubic metres (3.5×10^{12} cubic feet). INA estimates, on the other hand, are 50% lower than those supplied by Eni.[188] Oil was discovered in the Northern Adriatic, at a depth of approximately 5400 metres (17700 feet). The discovery was assessed as not viable because of its location, depth and quality.[189] Those gas and oil reserves are a part of the Po basin Province of Northern Italy and the Northern Mediterranean Sea.[190]

In 2000s, investigation works aimed at discovering gas and oil reserves in Middle Adriatic and Southern Adriatic basins was intensified, and by the end of the decade, oil and natural gas reserves were discovered Southeast of Bari and Brindisi—Rovesti and Giove oil discoveries. Surveys indicate reserves of 3 billion barrels of oil in place and 2 trillion cubic feet of gas in place.[191] The discovery was followed by start of further surveys off Croatian coast.[192] In January 2012, INA commenced prospecting for oil off Dubrovnik, marking resumption of oil exploration along the Eastern Adriatic coast after surveys commenced in late 1980s around the island of Brač were cancelled because of the war. Montenegro is also expected to look for oil off its coast.[193] As of January 2012, only 200 exploration wells were sunk off Croatian coast, with all but 30 executed in the Northern Adriatic basin.[194]

See also

- Mediterranean Sea

References

[1] Cushman-Roisin, Gačić, Poulain; pp. 1–2

[2] "Limits of Oceans and Seas, 3rd edition" (http://www.iho-ohi.net/iho_pubs/standard/S-23/S23_1953.pdf). International Hydrographic Organization. 1953. . Retrieved 7 February 2010.

[3] Wolfgang Ludwig et al. (2009). "River discharges of water and nutrients to the Mediterranean and Black Sea: Major drivers for ecosystem changes during past and future decades?" (http://www.sesame-ip.eu/doc/prooce_wl_inpress.pdf) (PDF). SESAME. . Retrieved 27 January 2012.

[4] "Drainage Basin of the Mediterranean Sea" (http://www.unece.org/fileadmin/DAM/env/water/publications/assessment/English/ J_PartIV_Chapter6_En.pdf) (PDF). United Nations Economic Commission for Europe. . Retrieved 27 January 2012.

[5] Nicholas Pinter, Gyula Grenerczy, John Weber (2006). *The Adria microplate: GPS geodesy, tectonics and hazards* (http://books.google.hr/ books?id=WBBJsPJ6C6QC). Springer. pp. 224-225. ISBN 9781402042348. . Retrieved 5 February 2012.

[6] Dražen Tutić; Miljenko Lapaine (2011). "Cartography in Croatia 2007–2011 - National Report to the ICA" (http://icaci.org/documents/ national_reports/2007-2011/Croatia.pdf) (PDF). International Cartographic Association. . Retrieved 5 February 2012.

[7] Gerald Henry Blake; Duško Topalović; Clive H. Schofield (1996). *The maritime boundaries of the Adriatic Sea* (http://books.google.hr/ books?id=lLVFW0an7NUC). IBRU. pp. 1-5. ISBN 9781897643228. . Retrieved 26 January 2012.

[8] *Length of the state border* (http://www.stat.si/letopis/2011/01-11.pdf). "Territory and climate" (PDF). *Statistical Yearbook 2011* (Statistical Office of the Republic of Slovenia) (50): 38. 2011. ISSN 1318-5403. . Retrieved 2 February 2012.

[9] Josip Faričić; Vera Graovac; Anica Čuka (June 2010). "Croatian small islands – residential and/or leisure area" (http://hrcak.srce.hr/index. php?show=clanak&id_clanak_jezik=84063). *Geoadria* (University of Zadar) 15 (1): 145-185. . Retrieved 28 January 2012.

[10] "Geographical and Meteorological Data" (http://www.dzs.hr/Hrv_Eng/ljetopis/2011/SLJH2011.pdf) (PDF). *2011 Statistical Yearbook of the Republic of Croatia* (Croatian Bureau of Statistics) 43: 41. December 2011. ISSN 1333-3305. . Retrieved 28 January 2012.

[11] Duncan Garwood (2009). *Mediterranean Europe*. Lonely Planet. p. 481. ISBN 9781741048568.

[12] Holly Hughes, Alexis Lipsitz Flippin, Sylvie Murphy, Julie Duchaine (2010). *Frommer's 500 Extraordinary Islands* (http://books.google. hr/books?id=ajhU1AAacUsC). John Wiley & Sons. p. 58. ISBN 9780470500705. . Retrieved 29 January 2012.

[13] Google, Inc. *Google Maps – Cape Kephali, Corfu, Greece - the southernmost point of the Adriatic Sea* (http://maps.google.com/ maps?q=cape+kephali,+greece&hl=en&ie=UTF8&ll=39.838068,19.25354&spn=1.090324,2.705383&sll=39.712469,19.647675& sspn=0.273083,0.676346&oq=Cape+&hnear=Ãkra+KavokefalÃ-&t=m&z=9) (Map). Cartography by Google, Inc. . Retrieved 29 January 2012.

[14] Cushman-Roisin, Gačić, Poulain; pp. 2–6

[15] Piero Mannini, Fabio Massa, Nicoletta Milone. "Adriatic Sea Fisheries: outline of some main facts" (http://www.faoadriamed.org/pdf/ publications/td13/mmm-td-13.pdf) (PDF). FAO AdriaMed. . Retrieved 29 January 2012.

[16] Cushman-Roisin, Gačić, Poulain; p. 218

[17] Introduction to *Previsioni di Marea nell'Alto Adriatico* (in Italian), Venice, issue 29 of year 29, by Stefano Fracon

[18] "Venice Municipality - Tide Monitoring and Forecast Center - Weather and sea parameters and their statistics" (http://www.comune. venezia.it/flex/cm/pages/ServeBLOB.php/L/IT/IDPagina/3045) (in Italian). .

[19] "Flood waters drench city of Venice" (http://news.bbc.co.uk/2/hi/8428781.stm). BBC News. 23 December 2009. . Retrieved 26 January 2012.

[20] "U Sloveniji more poplavilo obalu [Sea floods shore in Slovenia]" (http://dnevnik.hr/vijesti/svijet/ u-zasnijezenoj-sloveniji-raste-vodostaj-rijeka-more-poplavilo-obalu.html) (in Croatian). Nova TV (Croatia). 9 November 2010. . Retrieved 26 January 2012.

[21] "Zadar: Zbog velike plime more poplavilo obalu [Zadar: Sea floods shore because of high tide]" (http://dnevnik.hr/vijesti/hrvatska/ zadar-zbog-plime-more-poplavilo-rivu.html) (in Croatian). Nova TV (Croatia). 25 December 2009. . Retrieved 26 January 2012.

[22] "Jugo i niski tlak: More poplavilo šibensku rivu" (http://www.index.hr/vijesti/clanak/jugo-i-niski-tlak-more-poplavilo-sibensku-rivu/ 526311.aspx) (in Croatian). index.hr. 1 December 2010. . Retrieved 26 January 2012.

[23] Zdenko Franić; Branko Petrinec (September 2006). "Marine Radioecology and Waste Management in the Adriatic" (http://hrcak.srce.hr/ index.php?show=clanak&id_clanak_jezik=7507&lang=en). *Archives of Industrial Hygiene and Toxicology* (Institute for Medical Research

and Occupational Health) **57** (3): 347-352. ISSN 0004-1254. . Retrieved 04 February 2012.

[24] Alain Saliot (2005). *The Mediterranean Sea* (http://books.google.hr/books?id=UFHNGHADv5IC). Birkhäuser. p. 6. ISBN 9783540250180. . Retrieved 27 January 2012.

[25] Klement Tockner, Urs Uehlinger, Christopher T. Robinson (2009). "1.6 Hydrology and Biogeochemistry" (http://books.google.hr/ books?id=GDmX5XKkQCcC). *Rivers of Europe*. Academic Press. ISBN 9780123694492. . Retrieved 3 February 2012.

[26] Makoto Taniguchi et al. (2002). "Investigation of submarine groundwater discharge" (http://www.clas.ufl.edu/users/jbmartin/website/ Classes/Surface_Groundwater/Class 6/Taniguchi et al.2002 Hydrological processes review.pdf) (PDF). *Hydrological Processes* (John Wiley & Sons) **16**: 2115-2159. ISSN 1099-1085. . Retrieved 27 January 2012.

[27] Jože Žumer (2004). "Odkritje podmorskih termalnih izvirov [Discovery of submarine thermal springs]" (in Slovene). *Geografski Obzornik* (Association of the Geographical Societies of Slovenia) **51** (2): 11-17. ISSN 0016-7274.

[28] Giuseppe Colombo (1992). *Marine eutrophication and population dynamics* (http://books.google.com/books?id=vY650GD-OLkC). Olsen & Olsen. p. 380. ISBN 9788785215192. . Retrieved 28 January 2012.

[29] Cushman-Roisin, Gačić, Poulain; p. 145

[30] Artegiani et al. (August 1997). "The Adriatic Sea General Circulation. Part I: Air–Sea Interactions and Water Mass Structure" (http:// journals.ametsoc.org/doi/pdf/10.1175/1520-0485(1997)027<1492:TASGCP>2.0.CO;2) (PDF). *Journal of Physical Oceanography* (American Meteorological Society) **27**: 1492-1514. ISSN 0022-3670. . Retrieved 27 January 2012.

[31] Piero Lionello, Paola Malanotte-Rizzoli, R. Boscolo (2006). *Mediterranean climate variability* (http://books.google.hr/ books?id=JD8CqjuA4SAC). Elsevier. pp. 47-53. ISBN 9780444521705. . Retrieved 2 February 2012.

[32] "Led okovao svjetionik, u Tisnom smrznulo more [Ice covers a lighthouse, sea freezes at Tisno]" (http://www.rtl.hr/vijesti/novosti/ 24169/led-okovao-svjetionik-u-tisnom-smrznulo-more/) (in Croatian). RTL Televizija. 16 December 2010. . Retrieved 2 February 2012.

[33] "Mediterranean climate: Background information" (http://www.dsm.unisalento.it/webutenti/piero.lionello/public/book/mb_0_PL/ chapter0_V6.pdf) (PDF). University of Salento. . Retrieved 27 January 2012.

[34] Tomislav Šegota; Anita Filipčić (June 2003). "Köppenova podjela klima i hrvatsko nazivlje [Köppen climate classification and Croatian terminology]" (http://hrcak.srce.hr/9626) (in Croatian). *Geoadria* (University of Zadar) **8** (1): 17-37. . Retrieved 27 January 2012.

[35] Cushman-Roisin, Gačić, Poulain; pp. 6–8

[36] "World Weather Information Service" (http://www.worldweather.org/europe.htm). World Meteorological Organization. . Retrieved 27 January 2012.

[37] "Census of Population, Households and Dwellings 2011, First Results by Settlements" (http://www.dzs.hr/Hrv_Eng/publication/2011/ SI-1441.pdf) (in Croatian and English) (PDF). *Statistical Reports* (Zagreb: Croatian Bureau of Statistics) (1441). June 2011. ISSN 1332-0297. . Retrieved 30 June 2011.

[38] "Vista per singola area [Individual area review]" (http://demo.istat.it/pop2011/index.html) (in Italian). National Institute of Statistics (Italy). . Retrieved 30 January 2012.

[39] "Population and Housing Census in Albania" (http://census.al/Resources/Data/Census2011/Instat_print .pdf). Institute of Statistics of Albania. 2011. .

[40] Bogdan Sekulić; Ivan Sondi (December 1997). "Koliko je Jadran doista opterećen antropogenim i prirodnim unosom tvari? [To What Extent is the Adriatic Sea Actually Burdened with Men-induced and Natural Inflow of Substances]" (http://hrcak.srce.hr/index. php?show=clanak&id_clanak_jezik=95871) (in Croatian). *Hrvatski geografski glasnik* (Croatian Geographic Society) **59** (1). ISSN 1331-5854. . Retrieved 27 January 2012.

[41] "Facing Water Challenges in the Po River Basin, Italy:A WWDR3 Case Study" (http://waterwiki.net/index.php/ Facing_Water_Challenges_in_the_Po_River_Basin,_Italy:A_WWDR3_Case_Study). waterwiki.net. 2009. . Retrieved 6 April 2009.

[42] Raggi, Meri; Davide Ronchi; Laura Sardonini; Davide Viaggi (2007). "Po Basin Case study status report" (http://www.aquamoney. ecologic-events.de/sites/download/poit.pdf) (pdf). AquaMoney. . Retrieved 6 April 2009.

[43] *Technology: Venetians put barrage to the test against the Adriatic* (http://web.archive.org/web/20071011072114/http://media. newscientist.com/article/mg12216602.900-technology-venetians-put-barrage-to-the-test-against-theadriatic-.html). New Scientist magazine. 15 April 1989. Archived from the original (http://media.newscientist.com/article/mg12216602. 900-technology-venetians-put-barrage-to-the-test-against-theadriatic-.html) on 11 October 2007. . Retrieved 10 October 2007.

[44] "Venice's 1,500-year battle with the waves" (http://news.bbc.co.uk/2/hi/europe/3069305.stm). BBC News. 17 July 2003. . Retrieved 10 October 2007.

[45] "'Moses project' to secure future of Venice" (http://www.telegraph.co.uk/news/worldnews/europe/italy/3629387/ Moses-project-to-secure-future-of-Venice.html). Telegraph News. 11 January 2012. . Retrieved 11 January 2012.

[46] Maša Surić (June 2005). "Submerged Karst – Dead or Alive? Examples from the Eastern Adriatic Coast (Croatia)" (http://hrcak.srce.hr/ 9653). *Geoadria* (University of Zadar) **10** (1): 5-19. ISSN 1331-2294. . Retrieved 28 January 2012.

[47] "Mladen Juračić [Geology of the sea, Mediterranean and Adriatic]" (in Croatian). University of Zagreb, Faculty of Science, Geological Department.

[48] Nicholas Pinter (2006). *The Adria microplate: GPS geodesy, tectonics and hazards* (http://books.google.hr/books?id=z6h_2UnYIeEC). Springer. p. 352. ISBN 9781402042331. . Retrieved 28 January 2012.

[49] Sara Goudarzi (25 January 2008). "New Fault Found in Europe; May "Close Up" Adriatic Sea" (http://news.nationalgeographic.com/ news/2008/01/080125-europe-fault.html). National Geographic Society. . Retrieved 28 January 2012.

[50] F. Antonioli et al. (2007) (PDF). *Sea-level change during the Holocene in Sardinia and in the northeastern Adriatic (central Mediterranean Sea) from archaeological and geomorphological data* (http://people.rses.anu.edu.au/lambeck_k/pdf/265.pdf). Elsevier. pp. 2463-2486. ISSN 0277-3791. . Retrieved 4 February 2012.

[51] Mario Parise (2011). "Surface and subsurface karst geomorphology in the Murge (Apulia, Southern Italy)" (http://carsologica.zrc-sazu.si/downloads/401/Parise.pdf) (PDF). *Acta Carsologica* (Slovenian Academy of Sciences and Arts) **40** (1): 73-93. ISSN 0583-6050. . Retrieved 28 January 2012.

[52] Timi Ećimović, Elmar A. Stuhler, Marjan Vezjak (2000). *Anthology: SEM Institue for Cliamte Change* (http://books.google.hr/books?id=KOn0XjYNw_IC). Rainer Hampp Verlag. p. 115. ISBN 9783879885015. . Retrieved 3 February 2012.

[53] Siegfried Siegesmund (2008). *Tectonic aspects of the Alpine-Dinaride-Carpathian system* (http://books.google.hr/books?id=864MOs0qeuEC). Geological Society. pp. 146-149. ISBN 9781862392526. . Retrieved 3 February 2012.

[54] Eldridge M. Moores, Rhodes Whitmore Fairbridge (1997). *Encyclopedia of European and Asian regional geology* (http://books.google.hr/books?id=aYRup5mRcGsC). Springer. pp. 7-16. ISBN 9780412740404. . Retrieved 3 February 2012.

[55] Livio Vezzani, Andrea Festa, Francesca C. Ghisetti (2010). *Geology and tectonic evolution of the central-southern Apennines, Italy* (http://books.google.hr/books?id=qBNr2kdRhTcC). Geological Society of America. pp. 6 56. ISBN 9780813724690. . Retrieved 3 February 2012.

[56] McKinney, Frank Kenneth (2007). "Chapter 6: Pleistocene and Holocene Sediments" (http://books.google.hr/books/about/The_northern_Adriatic_ecosystem.html?id=TuxEf_Yn4YkC). *The northern Adriatic ecosystem: deep time in a shallow sea* (illustrated ed.). Columbia University Press. ISBN 9780231132428. . Retrieved 3 February 2012.

[57] J.P.M. Syvitski et al. (October, 2005). "Distributary channels and their impact on sediment dispersal" (http://www.sciencedirect.com/science?_ob=ArticleURL&_udi=B6V6M-4GWBWFT-2&_user=131115&_coverDate=11/15/2005&_rdoc=1&_fmt=high&_orig=search&_sort=d&_docanchor=&view=c&_searchStrId=1343366263&_rerunOrigin=google&_acct=C000009658&_version=1&_urlVersion=0&_userid=131115&md5=2f586a27a37212d31d722443c048f0aa). *Marine Geology* (Elsevier) **222-223** (15): 75–94. doi:10.1016/j.margeo.2005.06.030. . Retrieved 2010-05-21.

[58] "Geological processes in the Anthropocene: the Po River Delta" (http://www.theseusproject.eu/wiki/Geological_processes_in_the_Anthropocene:_the_Po_River_Delta). Theseus Project. . Retrieved 28 January 2012.

[59] Juan Antonio López Geta, ed (2003). *Coastal Aquifers Intrusion Technology: Mediterranean Countries* (http://books.google.hr/books?id=v_45Mb-OJIcC). Instituto Geológico y Minero de España. p. 210. ISBN 9788478404711. . Retrieved 3 February 2012.

[60] "Slovenian Sea" (http://www.vlada.si/en/about_slovenia/geography/the_slovenian_sea/). Government of Slovenia. . Retrieved 3 February 2012.

[61] "Rapid assessment of the Ecological Value of the Bojana-Buna Delta" (http://www.euronatur.org/uploads/media/Chapt_1-3_Rapid_assessment_of_the_Ecological_Value_of_the_Bojana-Buna_Delta_01.pdf) (PDF). EURONATUR. . Retrieved 3 February 2012.

[62] Jasmina Mužinić. "The Neretva Delta: Green Pearl of Coastal Croatia" (http://www.ncbi.nlm.nih.gov/pmc/articles/PMC2121601/). National Center for Biotechnology Information. . Retrieved 3 February 2012.

[63] Chemonics International Inc. (31 December 2000). "Biodiversity assessment for Croatia" (http://rmportal.net/library/content/118_croatia/view?searchterm=en). Natural Resources Management & Development Portal. . Retrieved 31 January 2012.

[64] Jasna Dolenc Koce, Barbara Vilhar, Borut Bohanec, Marina Dermastia (2003). "Genome size of Adriatic seagrasses" (http://www.vliz.be/imisdocs/publications/57225.pdf) (PDF). *Aquatic Botany* **77**: 17-25. ISSN 0304-3770. . Retrieved 30 January 2012.

[65] Lovrenc Lipej; Jakov Dulčić (2010). *Checklist of the Adriatic Sea Fishes* (http://books.google.hr/books/about/Checklist_of_the_Adriatic_Sea_Fishes.html?id=agwskgAACAAJ). Magnolia Press. ISBN 9781869775759. . Retrieved 30 January 2012.

[66] "The potential of Maritime Spatial Planning in the Mediterranean Sea - Case study report: The Adriatic Sea" (http://ec.europa.eu/maritimeaffairs/documentation/studies/documents/case_study_adriatic_sea_en.pdf) (PDF). European Union. 5 January 2011. . Retrieved 30 January 2012.

[67] "La Riserva [The reserve]" (http://www.riservamarinamiramare.it/riserva/index.htm) (in Italian). La Riserva Marina di Miramare. . Retrieved 31 January 2012.

[68] "The Marine Protected Area" (http://www.torredelcerrano.it/en/the-marine-protected-area.html). Marine Protected Area Torre Cerrano. . Retrieved 31 January 2012.

[69] "Riserva Marina delle Isole Tremiti [Tremiti Islands Marine Reserve]" (http://www.tremiti.eu/isole/riserva_marina/parco.aspx) (in Italian). tremiti.eu. . Retrieved 31 January 2012.

[70] "Protected Area" (http://www.parks.it/riserva.marina.torre.guaceto/Epar.php). Consorzio di Gestione di Torre Guaceto. . Retrieved 31 January 2012.

[71] "The Annotated Ramsar List: Italy" (http://www.ramsar.org/cda/en/ramsar-pubs-annolist-anno-italy/main/ramsar/1-30-168^16565_4000_0__). Ramsar Convention. . Retrieved 3 February 2012.

[72] "General Info" (http://www.brijuni.hr/en/general_info). Brijuni national park. . Retrieved 31 January 2012.

[73] "Documents and reports" (http://www.brijuni.hr/en/about_us/documents_and_reports). Brijuni national park. . Retrieved 31 January 2012.

[74] "Limski kanal [Lim Canal]" (http://www.rovinj.hr/rovinj/rovinj/priroda/limski-kanal) (in Croatian). City of Rovinj. . Retrieved 31 January 2012.

[75] "Welcome back to the National Park Kornati" (http://www.kornati.hr/eng/index.asp#). Kornati national park. . Retrieved 31 January 2012.

[76] "Nature Park Telašćica" (http://www.telascica.hr/opci-podaci.php?lang=en). Telašćica nature park. . Retrieved 31 January 2012.

[77] "Opće informacije [General information]" (http://pp-lastovo.hr/hr/opce-informacije) (in Croatian). Lastovo nature park. . Retrieved 31 January 2012.

[78] "The Annotated Ramsar List: Croatia" (http://www.ramsar.org/cda/en/ramsar-pubs-annolist-anno-croatia/main/ramsar/ 1-30-168^16461_4000_0__). Ramsar Convention. . Retrieved 3 February 2012.

[79] "National Report on the Application of the Protocol Concerning Specially Protected Areas and Biological Diversity in the Mediterranean" (http://195.97.36.231/dbases/Meeting Documents (Word or WP)/2005/05 WG 268 (SPA NFP - Seville)/Annex III.pdf). United Nations Environment Programme. 30 June 2005. pp. 227–228. .

[80] "Naravni spomenik Jezeri v Fiesi [The Fiesa Lakes Natural Monument]" (http://www.zrsvn.si/sl/informacija.asp?id_meta_type=63& id_informacija=651) (in Slovene). Institute of the Republic of Slovenia for Nature Conservation. . Retrieved 4 February 2012.

[81] *Enlarged protected areas of nature - natural parks, 30 June 2011* (http://www.stat.si/letopis/2011/01-11.pdf). "Territory and climate" (PDF). *Statistical Yearbook 2011* (Statistical Office of the Republic of Slovenia) (50): 38. 2011. ISSN 1318-5403. . Retrieved 4 February 2012.

[82] Pipan, Primož. "Sečoveljske soline [Sečovlje Salt Pans]" (http://www.dedi.si/dediscina/430-secoveljske-soline). In Šmid Hribar, Mateja. Torkar, Gregor. Golež, Mateja. Podjed, Dan. Drago Kladnik, Drago. Erhartič, Bojan. Pavlin, Primož. Jerele, Ines. (in Slovene). *Enciklopedija naravne in kulturne dediščine na Slovenskem — DEDI*. . Retrieved 3 February 2012.

[83] *Report of the Republic of Slovenia on the implementation of the Agreement in the period 2005-2007* (http://www.unep-aewa.org/ meetings/en/mop/mop4_docs/national_reports/pdf/slovenia2008.pdf). Ministry of the Environment and Spatial Planning, Republic of Slovenia. African-Eurasian Waterbird Agreement Secretariat. 2008. .

[84] Bratina Jurkovič, Nataša (April 2011). "Saltpans of Strunjan, Slovenia — proposal" (http://www.pap-thecoastcentre.org/pdfs/ Synthesis_Report_web.pdf). *Landscape Management Methodologies: Synthesis report of thematic studies*. Regional Activity Centre for the Priority Actions Programme. pp. 286–291. .

[85] "Osebna izkaznica [Identity Card]" (http://skocjanski-zatok.org/rezervat/osebna-izkaznica/) (in Slovene). Bird Watching and Bird Study Society of Slovenia. . Retrieved 4 February 2012.

[86] Turk, Robert. Odorico, Roberto (2009). "Marine protected areas in the Northern Adriatic" (http://www.zrsvn.si/dokumenti/63/2/2009/ Turk_Odorico_1585.pdf). *Varstvo narave* (Institute of the Republic of Slovenia for Nature Conservation) (22): 40. .

[87] Krivograd Klemenčič, Aleksandra. Vrhovšek, Danijel. Smolar-Žvanut, Nataša (31 March 2007). [hrcak.srce.hr/file/20762 "Microplanktonic and Microbenthic Algal Assemblages in the Coastal Brackish Lake Fiesa and the Dragonja Estuary (Slovenia)"]. *Natura Croatica* (Croatian Natural History Museum) **16** (1). hrcak.srce.hr/file/20762.

[88] "VENDIM PËR SHPALLJEN "PARK KOMBËTAR" TË EKOSISTEMIT NATYROR DETAR PRANË GADISHULLIT TË KARABURUNIT DHE ISHULLIT TË SAZANIT [Decision to declare "NATIONAL PARK" TO THE MARITIME natural ecosystem Karaburun Peninsula and the island of Sazan]" (http://lajme.shqiperia.com/lajme/artikull/iden/1046857853/titulli/ VENDIM-PER-SHPALLJEN-PARK-KOMBETAR-TE-EKOSISTEMIT-NATYROR-DETAR-PRANE-GADISHULLIT-TE--KARABURUNIT-DHE-ISHULL (in Albanian). shqiperia.com. . Retrieved 31 January 2012.

[89] "The Annotated Ramsar List: Albania" (http://www.ramsar.org/cda/en/ramsar-pubs-annolist-annotated-ramsar-17050/main/ramsar/ 1-30-168^17050_4000_0__). Ramsar Convention. . Retrieved 3 February 2012.

[90] Adrian Room (2006). *Placenames of the world* (http://books.google.hr/books?id=M1JIPAN-eJ4C). McFarland & Company. p. 20. ISBN 9780786422487. . Retrieved 26 January 2012.

[91] *The British Critic, volume 40* (http://books.google.hr/books?id=EQEwAAAAYAAJ). F. and C. Rivington. 1812. p. 504. . Retrieved 26 January 2012.

[92] Augustin Calmet; Charles Taylor (1830). *Calmet's dictionary of the Holy Bible* (http://books.google.hr/books?id=LOhSAAAAYAAJ). pp. 53-54. . Retrieved 26 January 2012.

[93] Charles Anthon (2005). *A Classical Dictionary: Containing the Principle Proper Names Mentioned in Ancient Authors* (http://books. google.hr/books?id=Wn949Sr-POkC). Kessinger Publishing. p. 20. ISBN 9781417976546. . Retrieved 26 January 2012.

[94] Cornelius Tacitus (1853). Charles Anthon. ed. *The Germania and Agricola, and also selections from the Annals, of Tacitus* (http://books. google.hr/books?id=NZgMAQAAMAAJ). Harper. p. 380. . Retrieved 26 January 2012.

[95] Barry Cunliffe (2008). *Europe Between the Oceans* (http://books.google.hr/books?id=nHUEAQAAIAAJ). Yale University Press. pp. 115-116. ISBN 9780300119237. . Retrieved 31 January 2012.

[96] John Franklin Hall (1996). *Etruscan Italy* (http://books.google.hr/books?id=bUhT7i7XhOAC). Indiana University Press. pp. 2-14. ISBN 9780842523349. . Retrieved 31 January 2012.

[97] John Wilkes (1995). *The Illyrians* (http://books.google.hr/books?id=4Nv6SPRKqs8C). Wiley-Blackwell. p. 91-104. ISBN 9780631198079. . Retrieved 31 January 2012.

[98] Gocha R. Tsetskhladze (2008). *Greek colonisation* (http://books.google.hr/books?id=z3C9b4FvpEwC). BRILL. pp. 155-186. ISBN 9789004155763. . Retrieved 31 January 2012.

[99] Erich S. Gruen (1986). *The Hellenistic world and the coming of Rome* (http://books.google.hr/books?id=EkdCokrrp4gC). University of California Press. p. 359. ISBN 9780520057371. . Retrieved 31 January 2012.

[100] Paul the Deacon (1974). *History of the Lombards* (http://books.google.hr/books?id=zt3jZ69vJKQC). University of Pennsylvania Press. pp. 326-328. ISBN 9780812210798. . Retrieved 31 January 2012.

[101] Thomas S. Burns (1991). *A history of the Ostrogoths* (http://books.google.hr/books?id=dw3FEpOUrRkC). Indiana University Press. pp. 126-130. ISBN 9780253206008. . Retrieved 31 January 2012.

[102] Samuel Griswold Goodrich (1856). *A history of all nations, from the earliest periods to the present time* (http://books.google.hr/books?id=YU3TAAAAMAAJ). Miller, Orton & Mulligan. p. 773. . Retrieved 31 January 2012.

[103] Andrew Archibald Paton (1861). *Researches on the Danube and the Adriatic* (http://books.google.com/books?id=E_NBAAAAYAAJ). Trübner. pp. 218–219. . Retrieved 15 October 2011.

[104] *Great history of Venice* (http://books.google.hr/books?id=mQL_EzSne5oC). Giunti Editore. pp. 61-74. ISBN 9788844005917. . Retrieved 31 January 2012.

[105] Daniel Don Nanjira (2010). *African Foreign Policy and Diplomacy* (http://books.google.hr/books?id=LZuxGsXVPoMC). ABC-CLIO. pp. 188-190. ISBN 9780313379826. . Retrieved 31 January 2012.

[106] Vladimir Posavec (March 1998). "Povijesni zemljovidi i granice Hrvatske u Tomislavovo doba [Historical maps and borders of Croatia in age of Tomislav]" (http://hrcak.srce.hr/index.php?show=clanak&id_clanak_jezik=62779) (in Croatian). *Radovi Zavoda za hrvatsku povijest* **30** (1): 281–290. ISSN 0353-295X. . Retrieved 16 October 2011.

[107] Norwich, pp. 250–253

[108] E. Glenn Hinson (1995). *The church triumphant* (http://books.google.hr/books?id=cY1SymrAGeEC). Mercer University Press. pp. 296-298. ISBN 9780865544369. . Retrieved 31 January 2012.

[109] Gordon S. Brown (2003). *The Norman conquest of Southern Italy and Sicily* (http://books.google.hr/books?id=c_Pft6RqfYIC). McFarland. pp. 3-5. ISBN 9780786414727. . Retrieved 31 January 2012.

[110] Fremont-Barnes, Gregory (2007). *Encyclopedia of the Age of Political Revolutions and New Ideologies, 1760-1815: Volume 1* (http://books.google.hr/books/about/Encyclopedia_of_the_age_of_political_rev.html?id=6_2wkP4j-EsC&redir_esc=y). Greenwood. p. 495. ISBN 978-0313334467. .

[111] Ladislav Heka (October 2008). "Hrvatsko-ugarski odnosi od sredinjega vijeka do nagodbe iz 1868. s posebnim osvrtom na pitanja Slavonije [Croatian-Hungarian relations from the Middle Ages to the Compromise of 1868, with a special survey of the Slavonian issue]" (http://hrcak.srce.hr/index.php?show=clanak&id_clanak_jezik=68144) (in Croatian). *Scrinia Slavonica* (Hrvatski institut za povijest – Podružnica za povijest Slavonije, Srijema i Baranje) **8** (1): 152–173. ISSN 1332-4853. . Retrieved 16 October 2011.

[112] Janet Sethre (2003). *The souls of Venice* (http://books.google.hr/books?id=5M7nwV0DvBYC). McFarland. p. 43-54. ISBN 9780786415731. . Retrieved 31 January 2012.

[113] Fernand Braudel (1992). *Civilization and Capitalism, 15th-18th Century* (http://books.google.hr/books?id=xMZI2QEer9QC). University of California Press. pp. 118-119. ISBN 9780520081161. . Retrieved 31 January 2012.

[114] Stanford J. Shaw (1976). *History of the Ottoman Empire and Modern Turkey* (http://books.google.hr/books?id=Xd422lS6ezgC). Cambridge University Press. p. 48. ISBN 9780521291637. . Retrieved 31 January 2012.

[115] Elisabeth Crouzet-Pavan; Lydia G. Cochrane (2005). *Venice Triumphant* (http://books.google.hr/books?id=B7gzrJDlUv8C). JHU Press. pp. 69-82. ISBN 9780801881893. . Retrieved 31 January 2012.

[116] Robert Browning (1992). *The Byzantine Empire* (http://books.google.hr/books?id=qp8ocRg7r2sC). CUA Press. p. 133. ISBN 9780813207544. . Retrieved 31 January 2012.

[117] Reinert, Stephen W. (2002). "Fragmentation (1204–1453)" (http://books.google.com/?id=Z6-kHUyyUIsC). In Cyril Mango. *The Oxford History of Byzantium*. Oxford: Oxford University Press. p. 270. ISBN 0198140983. .

[118] Richard C. Frucht (2005). *Eastern Europe: An Introduction to the People, Lands, and Culture* (http://books.google.hr/books?id=lVBB1a0rC70C). ABC-CLIO. pp. 422-423. ISBN 9781576078006. . Retrieved 18 October 2011.

[119] "Povijest saborovanja [History of parliamentarism]" (http://www.sabor.hr/Default.aspx?sec=404) (in Croatian). Sabor. . Retrieved 18 October 2010.

[120] G. W. Prothero et al. (1934). *The Cambridge modern history* (http://books.google.hr/books?id=u6w8AAAAIAAJ). CUP Archive. p. 495. . Retrieved 31 January 2012.

[121] Margaret L. King (2003). *The Renaissance in Europe* (http://books.google.hr/books?id=irQ5DKF69HwC). Laurence King Publishing. p. 202. ISBN 9781856693745. . Retrieved 31 January 2011.

[122] Charles Ingrao; Nikola Samardžić; Jovan Pesalj (2011). *The Peace of Passarowitz, 1718* (http://books.google.hr/books?id=T3Sg_1wR4poC). Purdue University Press. p. 66. ISBN 9781557535948. . Retrieved 1 February 2012.

[123] John Jeffries Martin; Dennis Romano (2002). *Venice reconsidered* (http://books.google.hr/books?id=PJEhbcRSELwC). JHU Press. p. 219. ISBN 9780801873089. . Retrieved 1 February 2012.

[124] H. Morse Stephens (2010). *Europe* (http://books.google.hr/books?id=PVh4b1_YmskC). Forgotten Books. pp. 192, 245. ISBN 9781440062179. . Retrieved 1 February 2012.

[125] Philippe Levillain (2002). *The Papacy: Gaius-Proxies* (http://books.google.hr/books?id=7VDcmDeLuV4C). Routledge. p. 1103. ISBN 9780415922302. . Retrieved 1 February 2012.

[126] Alexander I. Grab (2003). *Napoleon and the transformation of Europe* (http://books.google.hr/books?id=aW3aMxd5vrkC). Palgrave Macmillan. pp. 188-194. ISBN 9780333682746. . Retrieved 1 February 2012.

[127] James Henderson (1994). *The frigates* (http://books.google.hr/books?id=TH_fAAAAMAAJ). Leo Cooper. p. 112. ISBN 9780850524321. . Retrieved 1 February 2012.

[128] Patrick Richard Carstens; Timothy L Sanford (2011). *Searching for the Forgotten War - 1812 Canada* (http://books.google.hr/books?id=uybAm7NaF5YC). Xlibris Corporation. p. 425. ISBN 9781453588901. . Retrieved 1 February 2012.

[129] William M. James; Andrew D. Lambert (2002). *The Naval History of Great Britain, Volume 6, 1811–1827* (http://books.google.hr/ books?id=BH1nAAAAMAAJ). Conway Maritime. p. 180. ISBN 9780851779102. . Retrieved 1 February 2012.

[130] Harold Nicolson; Sir Harold Nicolson (2000). *The Congress of Vienna* (http://books.google.hr/books?id=ZTC3IWC_py8C). Grove Press. pp. 180, 226. ISBN 9780802137449. . Retrieved 1 February 2012.

[131] Rosita Rindler Schjerve (2003). *Diglossia and power* (http://books.google.hr/books?id=JOTFCTYSjicC). Walter de Gruyter. p. 200. ISBN 9783110176544. . Retrieved 1 February 2012.

[132] Roland Sarti (2004). *Italy* (http://books.google.hr/books?id=xhoLorFC1iwC). Infobase Publishing. p. 601. ISBN 9780816045228. . Retrieved 1 February 2012.

[133] Thomas F. X. Noble, Barry S. Strauss, Duane J. Osheim, Kristen B. Neuschel, Elinor Ann Accampo (2010). *Western Civilization: Beyond Boundaries* (http://books.google.hr/books?id=e5b57HtZH0kC). Cengage Learning. pp. 619-622. ISBN 9781424069620. . Retrieved 1 February 2012.

[134] Luciano Monzali (2009). *The Italians of Dalmatia* (http://books.google.hr/books?id=kMXURN7sxh4C). University of Toronto Press. pp. 73-76. ISBN 9780802096210. . Retrieved 1 February 2012.

[135] Sima M. Ćirković (2004). *The Serbs* (http://books.google.hr/books?id=sJJWEnXTHdIC). John Wiley & Sons. p. 225. ISBN 9780631204718. . Retrieved 1 February 2012.

[136] Spencer Tucker (2009). *A global chronology of conflict* (http://books.google.hr/books?id=h5_tSnygvbIC). ABC-CLIO. p. 1553. ISBN 9781851096671. . Retrieved 1 February 2012.

[137] Spencer C. Tucker, ed (2005). *World War I* (http://books.google.hr/books?id=2YqjfHLyyj8C). ABC-CLIO. p. 39. ISBN 9781851094202. . Retrieved 1 February 2012.

[138] Spencer Tucker; Laura Matysek Wood (1996). *The European powers in the First World War* (http://books.google.hr/ books?id=EHI3PCjDtsUC). Taylor & Francis. p. 440. ISBN 9780815303992. . Retrieved 1 February 2012.

[139] Alan Palmer (2000). *Victory 1918* (http://books.google.hr/books?id=ke3EOOUq3gwC). Grove Press. p. 298. ISBN 9780802137876. . Retrieved 1 February 2012.

[140] Tomasevich, pp. 4–16

[141] H. James Burgwyn (1997). *Italian foreign policy in the interwar period, 1918-1940* (http://books.google.hr/ books?id=PNHxISN-dmQC). Greenwood Publishing Group. pp. 4-16. ISBN 9780275948771. . Retrieved 1 February 2012.

[142] Stephen J. Lee (2003). *Europe, 1890-1945* (http://books.google.hr/books?id=Fln44TTgDLIC). Routledge. p. 318. . Retrieved 1 February 2012.

[143] Peter R. D'Agostino (2004). *Rome in America* (http://books.google.hr/books?id=U9wNxgCo7ycC). Univ of North Carolina Press. pp. 127-128. ISBN 9780807855157. . Retrieved 1 February 2012.

[144] Frederick Bernard Singleton (1985). *A short history of the Yugoslav peoples* (http://books.google.hr/books?id=qTLSZ3ucaZMC). Cambridge University Press. pp. 135-137. ISBN 9780521274852. . Retrieved 1 February 2012.

[145] Tomasevich, pp. 130–139, 233–234

[146] Matjaž Klemenčič (2004). *The former Yugoslavia's diverse peoples* (http://books.google.hr/books?id=ORSMBFwjAKcC). ABC-CLIO. pp. 198-202. ISBN 9781576072943. . Retrieved 1 February 2012.

[147] John J. Navone (1996). *The land and the spirit of Italy* (http://books.google.hr/books?id=TznfHevgsEcC). Legas / Gaetano Cipolla. pp. 141-142. ISBN 9781881901129. . Retrieved 1 February 2012.

[148] "History" (http://www.nato.int/history/index.html). NATO. . Retrieved 1 February 2012.

[149] Norman Polmar; Jurrien Noot (1991). *Submarines of the Russian and Soviet navies, 1718-1990* (http://books.google.hr/ books?id=7cDN8q2RHGMC). Naval Institute Press. pp. 169-170. ISBN 9780870215704. . Retrieved 1 February 2012.

[150] Chuck Sudetic (26 June 1991). "2 Yugoslav States Vote Independence to Press Demands" (http://www.nytimes.com/1991/06/26/ world/2-yugoslav-states-vote-independence-to-press-demands.html?ref=croatia). *The New York Times*. . Retrieved 1 February 2012.

[151] "Bosnia-Hercegovina timeline" (http://news.bbc.co.uk/2/hi/europe/1066981.stm). BBC News. . Retrieved 1 February 2012.

[152] "Timeline: Montenegro" (http://news.bbc.co.uk/2/hi/europe/country_profiles/5075632.stm). BBC News. . Retrieved 1 February 2012.

[153] Brendan O'Shea (2005). *The modern Yugoslav conflict 1991-1995* (http://books.google.hr/books?id=KoHQEEDzL5AC). Routledge. pp. 21-25. ISBN 9780415357050. . Retrieved 1 February 2012.

[154] Stjepan Bernardić (15 November 2004). "Admiral Letica je naredio: Raspali! [Admiral Letica ordered: Fire!]" (http://arhiv. slobodnadalmacija.hr/20041115/temedana01.asp) (in Croatian). *Slobodna Dalmacija*. . Retrieved 1 February 2012.

[155] John Pike. "Operation Maritime Guard" (http://www.globalsecurity.org/military/ops/maritime_guard.htm). Globalsecurity.org. . Retrieved 1 February 2012.

[156] Roger Cohen (27 November 1994). "NATO and the UN quarrel in Bosnia as Serbs press on" (http://www.nytimes.com/1994/11/27/ world/nato-and-the-un-quarrel-in-bosnia-as-serbs-press-on.html). *The New York Times*. . Retrieved 1 February 2012.

[157] "Heart of Belgrade bombed" (http://news.bbc.co.uk/2/hi/europe/310748.stm). BBC News. 3 April 1999. . Retrieved 1 February 2012.

[158] Gerald Henry Blake, Duško Topalović, Clive H. Schofield (1996). *The maritime boundaries of the Adriatic Sea* (http://books.google.hr/ books?id=lLVFW0an7NUC). IBRU. pp. 11-13. ISBN 9781897643228. . Retrieved 1 February 2012.

[159] Mladen Klemenčić; Duško Topalović (December 2009). "The maritime boundaries of the Adriatic Sea" (http://hrcak.srce.hr/index. php?show=clanak&id_clanak_jezik=71443). *Geoadria* (University of Zadar) 14 (2): 311-324. ISSN 1331-2294. . Retrieved 1 February 2012.

[160] Robert Bajruši (6 December 2007). "ZERP je nepotrebna avantura [ZERP is a needless adventure]" (http://www.nacional.hr/clanak/
       40630/zerp-je-nepotrebna-avantura) (in Croatian). Nacional (weekly). . Retrieved 2 February 2012.
[161] "About the Adriatic Euroregion" (http://www.adriaticeuroregion.org/index.php?option=com_content&view=article&id=68&
       Itemid=53&lang=en). Adriatic Euroregion. . Retrieved 5 February 2012.
[162] Piero Mannini, Fabio Massa, Nicoletta Milone. "Adriatic Sea Fisheries: outline of some main facts" (http://www.faoadriamed.org/pdf/
       publications/td13/mmm-td-13.pdf) (PDF). FAO AdriaMed. . Retrieved 30 January 2012.
[163] "Fishery and Aquaculture Country Profiles - Italy" (http://www.fao.org/fishery/countrysector/FI-CP_IT/en). Food and Agriculture
       Organization. . Retrieved 30 January 2012.
[164] "Fishery Country Profiles - Croatia" (ftp://ftp.fao.org/FI/DOCUMENT/fcp/en/FI_CP_HR.pdf) (PDF). Food and Agriculture
       Organization. . Retrieved 30 January 2012.
[165] "Country Fishery Information" (http://www.faoadriamed.org/html/country_p/AllCProfile.html). FAO AdriaMed. . Retrieved 30
       January 2012.
[166] "FEE is..." (http://www.fee-international.org/en). Fee-international.org. . Retrieved 30 January 2012.
[167] "Blue Flag Beaches and Marinas" (http://www.blueflag.org/Menu/Awarded+sites). Foundation for Environmental Education. .
       Retrieved 30 January 2012.
[168] "Tourism" (http://www.monstat.org/eng/page.php?id=43&pageid=43). Statistical office of Montenegro. . Retrieved 30 January 2012.
[169] Siniša Horak, PhD et al. (November 2011). "Glavni plan i strategija razvoja turizma Republike Hrvatske [General plan and strategy of
       development of tourism in the Republic of Croatia]" (http://www.mint.hr/UserDocsImages/Izvj-04_SRTRH.pdf) (in Croatian) (PDF).
       Institute for Tourism (Croatia). . Retrieved 30 January 2012.
[170] "Croatia - Key Facts at a Glance" (http://www.wttc.org/research/economic-impact-research/country-reports/c/croatia/). World Travel
       and Tourism Council. . Retrieved 30 January 2012.
[171] "Montenegro - Key Facts at a Glance" (http://www.wttc.org/research/economic-impact-research/country-reports/m/montenegro/).
       World Travel and Tourism Council. . Retrieved 30 January 2012.
[172] Fabio Quintiliani. "International tourism in the coastal regions of five Mediterranean countries" (http://www.rcfea.org/papers/WS1/
       Quintiliani.pdf) (PDF). The Rimini Centre for Economic Analysis. . Retrieved 30 January 2012.
[173] "Eurostat - Tourism" (http://epp.eurostat.ec.europa.eu/portal/page/portal/tourism/data/main_tables). Eurostat. . Retrieved 30 January
       2012.
[174] "INTERVJU: Načelnik općine Neum dr. Živko Matuško za BH. Privrednik [INTERVIEW: Municipal mayor dr. Živko Matuško for BH
       Privrednik]" (http://www.neum.ba/index.php/81-vijesti/126-intervju) (in Bosnian). Neum municipality. 17 October 2011. . Retrieved 30
       January 2012.
[175] "Pregled sektora turizma [A review of tourism sector]" (http://kfbih.com/udoc/10102011Izvozna_strategija_za_sektor_turizma.pdf) (in
       Bosnian) (PDF). Chamber of Economy of the Federation of Bosnia and Herzegovina. 2011. . Retrieved 30 January 2012.
[176] "Treguesit statistikorë të turizmit [Tourism statistical indicators]" (http://www.mtkrs.gov.al/web/
       Treguesit_statistikore_te_turizmit_30_1.php) (in Albanian). Ministry of Tourism, Cultural Affairs, Youth and Sports (Albania). . Retrieved
       30 January 2012.
[177] "Turizem [Tourism]" (http://www.stat.si/letopis/2011/25-11.pdf) (in Slovene). *Statistični letopis Republike Slovenije 2011 [Statistical
       Yearbook of the Republic of Slovenia 2011]*. Statistical Office of the Republic of Slovenia. 2012. ISSN 1318–5403. .
[178] "Turizem [Tourism]" (http://www.stat.si/doc/statinf/21-SI-O16-1101.pdf) (in Slovene). *Statistične informacije [Rapid Reports]*
       (Statistical Office of the Republic of Slovenia) (16): p. 5. 27 September 2011. .
[179] "Transport" (http://www.stat.si/letopis/2011/21-11.pdf) (PDF). Statistical Office of the Republic of Slovenia. 2011. . Retrieved 2
       February 2012.
[180] "Trasporti e telecomunicazioni [Transport and communications]" (http://www3.istat.it/dati/catalogo/20101119_00/PDF/cap19.pdf)
       (in Italian) (PDF). National Institute of Statistics (Italy). . Retrieved 2 February 2012.
[181] "TRAFFIC OF SHIPS, PASSENGERS AND GOODS BY HARBOUR MASTER'S OFFICES AND STATISTICAL PORTS, 2008"
       (http://www.dzs.hr/Hrv_Eng/ljetopis/2009/PDF/21-bind.pdf) (PDF). Croatian Bureau of Statistics. . Retrieved 1 February 2012.
[182] "Durrës, a good business choice" (http://www.ccidr.al/files/durres_a_good_business_choise.pdf) (PDF). Durrës' Chamber of
       Commerce and Industry - Albania. p. 14. . Retrieved 2 February 2012.
[183] "Slovenia's Luka Koper 2011 Cargo Throughput Up 11%" (http://limun.hr/en/main.aspx?id=779703). limun.hr. 17 January 2012. .
       Retrieved 2 February 2012.
[184] "Serbia eyes Montenegro's largest port" (http://www.b92.net/eng/news/business-article.php?yyyy=2009&mm=08&dd=26&
       nav_id=61375). B92. 26 August 2009. . Retrieved 2 February 2012.
[185] "Signed the founding of the NAPA in Trieste, Rijeka is expected to join in" (http://www.mmpi.hr/default.aspx?id=6598). Ministry of
       the Sea, Transport and Infrastructure (Croatia). 2 March 2010. . Retrieved 27 August 2011.
[186] "Port of Rijeka – Fifth Star of NAPA" (http://www.portsofnapa.com/index.php?t=news&id=6). North Adriatic Ports Association. 29
       November 2010. . Retrieved 27 August 2011.
[187] "Annamaria offshore oil rig starts trial run" (http://www.ina.hr/UserDocsImages/Ina_casopis/zima09-10/44-45 annamaria.pdf) (PDF).
       INA (company). 2009. . Retrieved 30 January 2012.
[188] "Natural gas reserves in the Adriatic may be up to 100 billion cubic meters" (http://www.limun.hr/en/main.aspx?id=312553). limun.hr.
       22 July 2008. . Retrieved 30 January 2012.

[189] Josip Sečen; Žarko Prnić (December 1996). "Istraživanje i proizvodnja ugljikovodika u Hrvatskoj [Prospecting and production of hydrocarbons in Croatia]" (http://hrcak.srce.hr/13551) (in Croatian). *Rudarsko-geološki-naftni zbornik* (University of Zagreb) **8** (1): 19-25. ISSN 0353-4529. . Retrieved 30 January 2012.

[190] "Petroleum Systems of the Po Basin Province of Northern Italy and the Northern Adriatic Sea" (http://pubs.usgs.gov/of/1999/ofr-99-0050/OF99-50M/occurrence.html). United States Geological Survey. . Retrieved 30 January 2012.

[191] "Northern Petroleum to expand exploration of Rovesti and Giove oil discoveries" (http://www.northpet.com/assets/uploads/Proactiveinvestors UK - Northern Petroleum to expand exploration of Rovesti and Giove oil discoveries.pdf) (PDF). Northern Petroleum. 28 July 2011. . Retrieved 30 January 2012.

[192] Sanjin Grandić; Slobodan Kolbah (2009). "New Commercial Oil Discovery at Rovesti Structure in South Adriatic and its Importance for Croatian Part of Adriatic Basin" (http://hrcak.srce.hr/46152). *Nafta* (Croatian Academy of Sciences and Arts, Scientific commussion for oil) **60** (2): 68-82. ISSN 0027-755X. . Retrieved 30 January 2012.

[193] Jasmina Mrvaljević (11 January 2012). "Naftu i plin vadit ćemo kod Dubrovnika [Oil and gas will be pumped near Dubrovnik]" (http://www.slobodnadalmacija.hr/Hrvatska/tabid/66/articleType/ArticleView/articleId/160871/Default.aspx) (in Croatian). *Slobodna Dalmacija*. . Retrieved 30 January 2012.

[194] Miho Dobrašin (3 January 2012). "Milanović poništio Kosoričin natječaj za istraživanje nafte i plina [Milanović cancels Kosor's oil and gas exploration tender procedure]" (http://www.business.hr/dogadjaji/milanovic-ponistio-kosoricin-natjecaj-za-istrazivanje-nafte-i-plina-106895) (in Croatian). Business.hr. . Retrieved 31 January 2012.

Bibliography

- Benoit Cushman-Roisin; Miroslav Gačić; Pierre-Marie Poulain (2001). *Physical oceanography of the Adriatic Sea* (http://books.google.hr/books?id=OFwkVgQNHlsC). Springer. ISBN 9781402002250. Retrieved 26 January 2012.

- John Julius Norwich (1997). *A short history of Byzantium* (http://books.google.hr/books?id=FkUmAQAAMAAJ). Knopf. ISBN 9780679772699. Retrieved 31 January 2012.

- Jozo Tomasevich (2001). *War and revolution in Yugoslavia, 1941-1945* (http://books.google.hr/books?id=fqUSGevFe5MC). Stanford University Press. ISBN 9780804736152. Retrieved 1 February 2012.

External links

- Region 5 - Western Africa, Mediterranean, Black Sea Nautical Charts (http://www.charts.noaa.gov/NGAViewer/Region_5_NGA_ViewerTable.shtml) from National Geospatial-Intelligence Agency
- Nautical Chart 54131 (Adriatic Sea) (http://www.charts.noaa.gov/NGAViewer/54131.shtml) from National Geospatial-Intelligence Agency

List_of_longest_bridges_in_the_world

This is a list of the world's bridges more than two kilometers long sorted by their full length above land or water. "Span" refers to the longest spans without any ground support.

*Note: There is no standard way to measure the total length of a bridge. Some bridges are measured from the beginning of the entrance ramp to the end of the exit ramp. Some are measured from shoreline to shoreline. Yet others are the length of the total construction involved in building the bridge. Since there is no standard, **no ranking of a bridge should be assumed because of its position in the list.** Additionally, numbers are merely estimates and measures in U.S. customary units (feet) may be imprecise due to conversion error.*

Name	Length metres (feet)	Span metres (feet)	Completed	Traffic	Country
Danyang–Kunshan Grand Bridge *Beijing–Shanghai High-Speed Railway* *Guinness: Longest bridge (any type), 2011*[1]	164800 m (ft)	80 m (260 ft)	2010 (Complete) 2011 (Open)	High-speed rail	People's Republic of China
Tianjin Grand Bridge[2] *Beijing–Shanghai High-Speed Railway*	113700 m (ft)	?	2010 (Complete) 2011 (Open)	High-speed rail	People's Republic of China
Weinan Weihe Grand Bridge *Zhengzhou-Xi'an High-Speed Railway*	79732 m (ft)	80 m (260 ft)[3]	2008 (Complete) 2010 (Open)	High-speed rail	People's Republic of China
Bang Na Expressway	54000 m (ft)	44 m (144 ft)	2000	Road	Thailand
Beijing Grand Bridge *Beijing–Shanghai High-Speed Railway*	48153 m (ft)	108 m (354 ft)	2010 (Complete) 2011 (Open)	High-speed rail	People's Republic of China
Jiaozhou Bay Bridge *Guinness: Longest bridge over water (aggregate), July 2011*[4]	42500 m (ft)[5][6]	260 m (850 ft)	2011	Road	People's Republic of China
Lake Pontchartrain Causeway *Guinness: Longest bridge over water (continuous), 1969*[7]	38442 m (ft)	46 m (151 ft)	1956 (SB) 1969 (NB)	Road	United States
Manchac Swamp bridge	36710 m (ft)	?	1970	Road	United States
Yangcun Bridge *Beijing-Tianjin Intercity Railway*	35812 m (ft)[8]	?	2007	High-speed rail	People's Republic of China
Hangzhou Bay Bridge	35673 m (ft)	448 m (1470 ft)	2007	Road	People's Republic of China
Runyang Bridge	35660 m (ft)[9]	1490 m (4890 ft)	2005	Road	People's Republic of China
Donghai Bridge	32500 m (ft)	400 m (1300 ft)	2005	Road	People's Republic of China
Shanghai Maglev line	29908 m (98123 ft)[10]	?	2003	Maglev	People's Republic of China
Atchafalaya Basin Bridge	29290 m (96100 ft)	?	1973	Highway	United States
Yanshi Bridge *Zhengzhou-Xi'an High-Speed Railway*	28543 m (93645 ft)[11]	?	2009	High-speed rail	People's Republic of China
Kanpur City Bypass (Kanpur over-bridge) (longest road bridge in Southern Asia)[12]	25000 m (82000 ft)	?	2003	Road	India

King Fahd Causeway	25000 m (82000 ft)[13]	?	1986	Road	Saudi Arabia and Bahrain
Jintang Bridge	26540 m (87070 ft)	?	2009	Road	People's Republic of China
Jinbin Light Rail No. 1 Bridge (Guanghualu – Babaocun) *Tianjin Binhai Mass Transit*	25800 m (84600 ft)	?	2003	Metro	People's Republic of China
Suvarnabhumi Airport Link	24500 m (80400 ft)	?	2010	Rail	Thailand
Chesapeake Bay Bridge-Tunnel (in Virginia)	24140 m (79200 ft)	?	1964 (NB) 1999 (SB)	Highway	United States
Liangshui River Bridge *Beijing-Tianjin Intercity Railway*	21563 m (70745 ft)[8]	?	2007	High-speed rail	People's Republic of China
Yongding New River Bridge *Beijing-Tianjin Intercity Railway*	21133 m (69334 ft)[14]	?	2007	High-speed rail	People's Republic of China
6th October Bridge	20500 m (67300 ft)	?	1996	Road	Egypt
C215 Viaduct[15] *Taiwan High Speed Rail*	20000 m (66000 ft)	?	2007	High-speed rail	Republic of China (Taiwan)
Incheon Bridge	18384 m (60315 ft)[16]	800 m (2600 ft)	2009	Road	South Korea
Cangzhou–Dezhou Grand Bridge[2] *Beijing–Shanghai High-Speed Railway*	18200 m (59700 ft)	128 m (420 ft)	2010 (Complete) 2011 (Open)	High-speed rail	People's Republic of China
Aérotrain Test Track *No longer in use*	18000 m (59000 ft)	?	1965	Rail (prototype)	France
Bonnet Carré Spillway bridge of I-10	17702 m (58077 ft)	?	1960s	Road	United States
Vasco da Gama Bridge *Longest bridge in Europe*	17185 m (56381 ft)	450 m (1480 ft)	1998	Highway	Portugal
Saigon - Trung Luong Skyway *North-South Expressway, Vietnam*	16000 m (52000 ft)	?	2010	Road	Vietnam
Cross Beijing Ring Roads Bridge *Beijing-Tianjin Intercity Railway*	15595 m (51165 ft)[8]	?	2007	High-speed Rail	People's Republic of China
Kama Bridge[17]	13967 m (45823 ft)	?	2002	Road	Russia
Penang Bridge	13500 m (44300 ft)	225 m (738 ft)	1985	Road	Malaysia
Kam Sheung Road-Tuen Mun viaduct (part of West Rail Line)	13400 m (44000 ft)	?	2003	Rail	Hong Kong
Wuppertal Schwebebahn	13300 m (43600 ft)	33 m (108 ft)	1903	Suspended monorail track	Germany
Rio-Niterói Bridge	13290 m (43600 ft)	300 m (980 ft)	1974	Road	Brazil
Bhumibol Bridge	13000 m (43000 ft)	398 m (1306 ft)	2006	Road	Thailand

New Ulyanovsk Bridge	12980 m (42590 ft)[18]	220 m (720 ft)	2009	Road and light metro	Russia
Confederation Bridge *longest bridge over ice (winter)*	12900 m (42300 ft)	250 m (820 ft) (43x)	1997	Highway	Canada
Jubilee Parkway	12875 m (42241 ft)	?	1978	Highway	United States
Novyi Saratovskiy Bridge[19]	12760 m (41860 ft)	1228 m (4029 ft)	2000	Road	Russia
Rudong Yangkou Yellow Sea Bridge[20]	12600 m (41300 ft)[21]	?	2008	Road	People's Republic of China
Emsland test facility	12000 m (39000 ft)	?	1985	Rail (Maglev)	Germany
Nanjing Qinhuai River Bridge *Beijing-Shanghai High-Speed Railway*	12000 m (39000 ft)[21]	?	2010	High-speed rail	People's Republic of China
Qingshuihe Bridge[22] *Qingzang Railway*	11700 m (38400 ft)	?	2006	Rail	People's Republic of China
Leziria Bridge[23]	11670 m (38290 ft)	133 m (436 ft)	2007		Portugal
Hyderabad (P.V. Expressway) Express way connecting Hyderabad to Hyderabad International Airport	11600 m (38100 ft)	?	2009	Road	India
San Mateo-Hayward Bridge	11265 m (36959 ft)	229 m (751 ft)[24]	1967	Highway	United States
Zhenjiang Beijing–Hangzhou Canal Bridge *Beijing-Shanghai High-Speed Railway*	11000 m (36000 ft)[21]	?	2010	High-speed rail	People's Republic of China
Seven Mile Bridge	10887 m (35719 ft)	41 m (135 ft)	1982	Highway	United States
Sunshine Skyway Bridge	10500 m (34400 ft)	366 m (1201 ft)	1987	Highway	United States
Third Mainland Bridge	10500 m (34400 ft)	?	1991	Road	Nigeria
Shandong-Henan Yellow River Bridge[25]	10282 m (33734 ft)	?	1985		People's Republic of China
Wuhu Yangtze River Bridge	10020 m (32870 ft)	312 m (1024 ft)	2000	Road & Rail	People's Republic of China
Hosur Road Elevated Expressway (Bangalore) Expressway connecting Downtown Bangalore to Electronics City on Hosur Road	9945 m (32628 ft)	?	2010	Highway	India
Shanghai Yangtze River Bridge	9970 m (32710 ft)	730 m (2400 ft)	2009	Road (& future Rail)	People's Republic of China
General W.K. Wilson Jr. Bridge	9786 m (32106 ft)	?	?	Highway	United States
Norfolk Southern Lake Pontchartrain Bridge	9300 m (30500 ft)	?	?	Railroad	United States

Nanjing Dashengguan Yangtze River Bridge	9273 m (30423 ft)[21]	336 m (1102 ft)	2010	High-speed Rail and Metro	People's Republic of China
Chacahoula Swamp Bridge	9005 m (29544 ft)	?	1995		United States
Twin Span bridge of Interstate 10	8851 m (29039 ft)	?	1962 (Original) 2009 (New WB) 2011 (New EB)	Highway	United States
Richmond-San Rafael Bridge	8851 m (29039 ft)	317 m (1040 ft)	1956	Highway	United States
General Rafael Urdaneta Bridge	8678 m (28471 ft)	235 m (771 ft)	1962	Road	Venezuela
Virginia Dare Memorial Bridge (in North Carolina)	8369 m (27457 ft)	?	2002	Highway	United States
Yangpu Bridge	8354 m (27408 ft)	602 m (1975 ft)	1993	Road	People's Republic of China
Xiasha Bridge[26]	8230 m (27000 ft)	232 m (761 ft)	1991		People's Republic of China
Sutong Bridge	8206 m (26923 ft)	1088 m (3570 ft)	2008	Road	People's Republic of China
Mackinac Bridge	8038 m (26371 ft)	1158 m (3799 ft)	1957	Highway	United States
Destrehan Swamp Freeway	7902 m (25925 ft)	?	1992	Highway	United States
Øresund Bridge	7845 m (25738 ft)	490 m (1610 ft)	1999	Highway & Railroad	Denmark/ Sweden
Maestri Bridge	7693 m (25240 ft)	11 m (36 ft)	1928	Road	United States
Jiujiang Yangtze River Bridge[27]	7675 m (25180 ft)	216 m (709 ft)	1992	Road & rail	People's Republic of China
James River Bridge	7425 m (24360 ft)	126 m (413 ft)	1983	Highway	United States
Gwangan Bridge	7420 m (24340 ft)	?	2002	Road	South Korea
Champlain Bridge (Montreal)	7414 m (24324 ft)	215 m (705 ft)	1967	Highway	Canada
Seohae Bridge[28]	7310 m (23980 ft)	470 m (1540 ft)	2000		South Korea
Volgograd Bridge	7110 m (23330 ft)	?	October 2009	?	Russia
Chesapeake Bay Bridge (in Maryland)	6946 m (22789 ft)	490 m (1610 ft)	1952, 1973	Highway	United States
Huey P. Long Bridge	7000 m (23000 ft)	?	1936	Highway & Railroad	United States
Great Belt Bridge (Eastern)	6790 m (22280 ft)	1624 m (5328 ft)	1998	Highway	Denmark
Nanjing Yangtze River Bridge	6772 m (22218 ft)	160 m (520 ft)	1968	Highway and Railroad	People's Republic of China

Great Belt Bridge (Western)	6611 m (21690 ft)	?	1998	Highway & Railroad	██ Denmark
Thanlwin Bridge (Mawlamyaing)	6589 m (21617 ft)	?	2005	Road, rail & pedestrian	Myanmar
St. George Island Bridge	6588 m (21614 ft)	366 m (1201 ft)	2004	Highway	United States
Astoria-Megler Bridge	6545 m (21473 ft)	375 m (1230 ft)	1966	Highway	United States
Öland bridge	6072 m (19921 ft)	130 m (430 ft)	1972	Highway	Sweden
Libertador General San Martín Bridge	5966 m (19573 ft)	220 m (720 ft)	1976	Road	Uruguay and Argentina
Hernando de Soto Bridge	5954 m (19534 ft)	274 m (899 ft)	1973	Highway	United States
Pulaski Skyway	5636 m (18491 ft)	168 m (551 ft)	1932	Highway	United States
Garden City Skyway	5633 m (18481 ft)	?	1963	Highway	Canada
Albemarle Sound Bridge[29]	5627 m (18461 ft)	?	1990		United States
Bandra-Worli Sea Link	5600 m (18400 ft)	250 m (820 ft)	2009	Road	India
M6 - Castle Bromwich to Gravelly Hill viaduct	5600 m (18400 ft)	-	1971	Road	United Kingdom
Mahatma Gandhi Setu	5575 m (18291 ft)	?	1982	Road	India
Shenzhen Bay Bridge	5545 m (18192 ft)	?	2006	Road	Hong Kong
Island Eastern Corridor (Causeway Bay to Quarry Bay section)	5500 m (18000 ft)		1983	Road	Hong Kong
Suramadu Bridge (cross Madura Strait)	5438 m (17841 ft)	434 m (1424 ft)	2009	Road	Indonesia
Dauphin Island Bridge	5430 m (17810 ft)	122 m (400 ft)	1982		United States
Xinkai River Bridge of Beijing-Tianjin Intercity Railway	5371 m (17621 ft)[14]	?	2007	High-speed Rail	People's Republic of China
Second Severn Crossing	5128 m (16824 ft)	456 m (1496 ft)	1996	Road	United Kingdom
Zeeland Bridge	5022 m (16476 ft)	95 m (312 ft)	1965	Road	Netherlands
Malir River Bridge	5000 m (16000 ft)	?	2009	Road	Pakistan
Candaba Viaduct[30]	5000 m (16000 ft)	?	2005		Philippines
Buckman Bridge	4968 m (16299 ft)	76 m (249 ft)	1970	Highway	United States

Name	Total length	Main span	Year	Type	Country
Tappan Zee Bridge	4881 m (16014 ft)	736 m (2415 ft)	1955	Highway	United States
Howard Frankland Bridge *II*	4846 m (15899 ft)	?	1991	Highway	United States
Wright Memorial Bridge	4828 m (15840 ft)	?	1930	Highway	United States
Bangabandhu Bridge	4800 m (15700 ft)	100 m (330 ft) (47x)	1998	Highway & Railroad	Bangladesh
Shenzhen Western Corridor Bridge	4770 m (15650 ft)	210 m (690 ft)	2007	Road	People's Republic of China
Vikramshila Setu	4700 m (15400 ft)	?	2001	Road	India
Vembanad Rail Bridge of Kerala (Cochin) Bridge connecting ICTT Vallarpadam to Shoranur-Ernakulam Railway Line	4620 m (15160 ft)	?	2010	Rail	India
Lindsay C. Warren Bridge[31]	4550 m (14930 ft)	?	1960		United States
Gandy Bridge *I*	4529 m (14859 ft)	?	1975	Highway	United States
Sault Ste. Marie International Bridge	4480 m (14700 ft)	?	1962	Highway	United States and Canada
Jingzhou Yangtze River Bridge[32]	4398 m (14429 ft)[33]	500 m (1600 ft)	2002		People's Republic of China
Aqua Bridge (Tokyo Bay Aqua-Line)	4384 m (14383 ft)	?	1997	Road	Japan
Ponte Salgueiro Maia	4300 m (14100 ft)	?	2000	Highway	Portugal
Bayside Bridge	4270 m (14010 ft)	?	1993	Highway	United States
Hochstraße Elbmarsch[34]	4258 m (13970 ft)	35 m (115 ft)	1974	Road	Germany
Commodore Barry Bridge	4240 m (13910 ft)	501 m (1644 ft)	1974	Highway	United States
Gandy Bridge *II*	4226 m (13865 ft)	?	1997	Highway	United States
Escambia Bay Bridge	4224 m (13858 ft)	?	2004 (new span)	Highway	United States
Greenville Bridge	4133 m (13560 ft)	420 m (1380 ft)	2007	Highway	United States
Rosario-Victoria Bridge	4098 m (13445 ft)	330 m (1080 ft)	2003	Road	Argentina
Crescent City Connection	4093 m (13428 ft)	480 m (1570 ft)	1958	Highway	United States
Arthur Ravenel, Jr. Bridge	4023 m (13199 ft)	471 m (1545 ft)	2005	Highway	United States
Fred Hartman Bridge	4000 m (13000 ft)	381 m (1250 ft)	1995	Highway	United States

Bridge	Total length	Longest span	Year	Type	Country
Zacatal Bridge	3982 m (13064 ft)	?	1994	Road	Mexico
Chris Smith Bridge	3954 m (12972 ft)	265 m (869 ft)	1973		United States
Köhlbrandbrücke	3940 m (12930 ft)	520 m (1710 ft)	1974	Road	Germany
Herbert C. Bonner Bridge[35]	3921 m (12864 ft)	?	1963		United States
Akashi-Kaikyō Bridge	3911 m (12831 ft)	1991 m (6532 ft)	1998	Highway	Japan
Lupu Bridge	3900 m (12800 ft)	550 m (1800 ft)	2003	Road	People's Republic of China
Suez Canal Bridge	3900 m (12800 ft)	440 m (1440 ft)	2001	Road	Egypt
Yuribey Bridge	3890 m (12760 ft)	110 m (360 ft)	2009	Rail	Russia
The First Kitakami River Bridge	3868 m (12690 ft)	?	1982	Rail	Japan
Ponte della Libertà	3850 m (12630 ft)	?	1846/1933	Road & rail	Italy
Queen Isabella Causeway[36]	3810 m (12500 ft)	?	1974	Road	United States
Mozambique Island Bridge[37]	3800 m (12500 ft)	?	1969	Road	Mozambique
Santhià Viaduct[38]	3782 m (12408 ft)	?	2006		Italy
Rodoferroviária Bridge	3770 m (12370 ft)	100 m (330 ft)	1998	Road & rail	Brazil
Sky Gate Bridge R[39]	3750 m (12300 ft)	?	1994	Road & Rail	Japan
Vinh Tuy Bridge	3690 m (12110 ft)		2009	Road	Vietnam
Dona Ana Bridge	3670 m (12040 ft)	80 m (260 ft)	1934	Rail	Mozambique
Walt Whitman Bridge	3652 m (11982 ft)	610 m (2000 ft)	1957	Highway	United States
Humen Pearl River Bridge	3618 m (11870 ft)	888 m (2913 ft)	1957	Road	People's Republic of China
Ayrton Senna Bridge	3607 m (11834 ft)	?	1998		Brazil
Fadalto Viaduct[40]	3567 m (11703 ft)	?	1990		Italy
Thang Long Bridge	3500 m (11500 ft)		1974	Road	Vietnam
Tay Rail Bridge	4500 m (14800 ft)	?	1887	Rail	United Kingdom

San Diego-Coronado Bridge	3407 m (11178 ft)	?	1969	Highway	United States
Lake Jesup Bridge	3379 m (11086 ft)	?	1993		United States
Saint-Nazaire Bridge	3356 m (11010 ft)	404 m (1325 ft)	1974	Road	France
Third Bridge	3300 m (10800 ft)	260 m (850 ft)	1989	Road	Brazil
Delaware Memorial Bridge *II*	3291 m (10797 ft)	655 m (2149 ft)	1968	Highway	United States
Delaware Memorial Bridge *I*	3281 m (10764 ft)	655 m (2149 ft)	1951	Highway	United States
Luling Bridge	3261 m (10699 ft)	376 m (1234 ft)	1983	Highway	United States
Dames Point Bridge	3245 m (10646 ft)	396 m (1299 ft)	1989	Highway	United States
Storstrøm Bridge	3199 m (10495 ft)	136 m (446 ft)	1937	Highway & Railroad	Denmark
Second Orinoco crossing (Orinoquia Bridge)	3156 m (10354 ft)	300 m (980 ft)	2006	Road & rail	Venezuela
San Francisco-Oakland Bay Bridge	3141 m (10305 ft)	18 m (59 ft)	1936	Highway	United States
Thanh Tri Bridge	3084 m (10118 ft)		2007	Road	Vietnam
Heishipu Bridge[41]	3068 m (10066 ft)	162 m (531 ft)	2004		People's Republic of China
Nehru Setu[42]	3065 m (10056 ft)	31 m (102 ft)	1900	Rail	India
Jawahar Setu	3061 m (10043 ft)	?	1965	Road	India
Talmadge Memorial Bridge	3060 m (10040 ft)	335 m (1099 ft)	1990	Highway	United States
Kolia Bhomora Setu	3015 m (9892 ft)	120 m (390 ft)	1987	Road	India
Naranarayana Bridge	2284 m (7493 ft)	125 m (410 ft)	1998	Road	India
Jiangyin Suspension Bridge	3000 m (9800 ft)	1385 m (4544 ft)	1999	Road	People's Republic of China
C310 Viaduct[43]	3000 m (9800 ft)	?	2007		Republic of China (Taiwan)
Re Island Bridge[44]	2927 m (9603 ft)	110 m (360 ft)	1988		France
Benjamin Franklin Bridge	2918 m (9573 ft)	533 m (1749 ft)	1926	Highway & Railroad	United States
Hiroshima Kaita Bridge	2900 m (9500 ft)	?	1990		Japan
Rio-Antirio bridge	2880 m (9450 ft)	1410 m (4630 ft)	2004	Road	Greece
Queen Elizabeth II Bridge (Dartford Crossing)	2872 m (9423 ft)	450 m (1480 ft)	1991	Road	United Kingdom

Rach Mieu Bridge	2868 m (9409 ft)	270 m (890 ft)	2009	Road	Vietnam
Oleron Bridge[45]	2862 m (9390 ft)	80 m (260 ft)	1966		France
Second Quiantang River Bridge[46]	2861 m (9386 ft)	80 m (260 ft)	1991		People's Republic of China
Rügenbrücke	2831 m (9288 ft)	583 m (1913 ft)	2007	Road	Germany
Big Obukhovsky Bridge	2824 m (9265 ft)	382 m (1253 ft)	2004	Road	Russia
Saratov Bridge	2804 m (9199 ft)	?	1965	Road	Russia
Giurgiu-Rousse Friendship Bridge	2800 m (9200 ft)	?	1954	Road & rail	Romania and Bulgaria
Hornibrook Bridge	2800 m (9200 ft)	?	1935 (Demolished Mid 2011)	Pedestrian & cyclist	Australia
Third Mainland Bridge	2800 m (9200 ft)	?	1988	Road	Nigeria
Can Tho Bridge	2750 m (9020 ft)	550 m (1800 ft)	2010	Road	Vietnam
Houghton Highway	2740 m (8990 ft)	?	1979	Road	Australia
Victoria Bridge	2790 m (9150 ft)	?	1859	Highway & Railroad	Canada
Old Godavari Bridge	2745 m (9006 ft)	?	1900		India
Golden Gate Bridge	2737 m (8980 ft)	1280 m (4200 ft)	1937	Highway	United States
New Godavari Bridge	2730 m (8960 ft)	?	1997	Road & rail	India
Laviolette Bridge	2707 m (8881 ft)	335 m (1099 ft)	1967	Highway	Canada
Jacques Cartier Bridge	2687 m (8816 ft)	334 m (1096 ft)	1930	Highway	Canada
Dumbarton Bridge	2621 m (8599 ft)	104 m (341 ft)	1982	Highway	United States
Banghwa Bridge[47]	2599 m (8527 ft)	180 m (590 ft)	2000		South Korea
Kremsbrücke Pressingberg[48]	2607 m (8553 ft)	?	1980	Road	Austria
Alex Fraser Bridge	2602 m (8537 ft)	465 m (1526 ft)	1986	Highway	Canada
Khabarovsk Bridge	2590 m (8500 ft)	?	1999	Road & rail	Russia
Betsy Ross Bridge	2586 m (8484 ft)	222 m (728 ft)	1976	Highway	United States
West Gate Bridge	2582 m (8471 ft)	336 m (1102 ft)	1978	Highway	Australia
Governador Nobre de Carvalho	2570 m (8430 ft)	?	1974	Road	Macau
Burlington Bay Skyway	2561 m (8402 ft)	150 m (490 ft)	1958	Highway	Canada
Richard I. Bong Memorial Bridge	2559 m (8396 ft)	?	1985	Highway	United States
Forth Bridge	2529 m (8297 ft)	521 m (1709 ft)	1890	Railroad	United Kingdom
Forth Road Bridge	2512 m (8241 ft)	1006 m (3301 ft)	1964	Road	United Kingdom

Name	Length	Span	Year	Type	Country
Sunshine Bridge	2510 m (8230 ft)	251 m (823 ft)	1964	Highway	United States
Penghu Trans-Oceanic Bridge	2494 m (8182 ft)	?	1970	Road	Taiwan
Thi Nai Bridge	2477 m (8127 ft)	15 m (49 ft)	2006	Road	Vietnam
Drežnik Viaduct[49]	2485 m (8153 ft)	70 m (230 ft)	2001		Croatia
Zilwaukee Bridge	2466 m (8091 ft)	119 m (390 ft)	1988	Road	United States
Millau Viaduct	2460 m (8070 ft)	342 m (1122 ft)	2004	Highway	France
Rama VIII Bridge	2450 m (8040 ft)	300 m (980 ft)	2002	Road & pedestrian	Thailand
Leo Frigo Memorial Bridge	2430 m (7970 ft)	137 m (449 ft)	1981	Road	United States
John A. Blatnik Bridge	2430 m (7970 ft)	?	1961	Highway	United States
Shantou Bay Bridge	2425 m (7956 ft)	452 m (1483 ft)	1995		People's Republic of China
Golden Ears Bridge	2410 m (7910 ft)	968 m (3176 ft)	2009	Highway	Canada
Bubiyan Bridge	2380 m (7810 ft)	54 m (177 ft)	1983		Kuwait
Kingston-Rhinecliff Bridge	2375 m (7792 ft)	244 m (801 ft)	1957	Highway	United States
Newburgh-Beacon Bridge	2374 m (7789 ft)	305 m (1001 ft)	1963	Highway	United States
Sidney Lanier Bridge	2371 m (7779 ft)	381 m (1250 ft)	2003	Highway	United States
General Artigas Bridge	2350 m (7710 ft)	334 m (1096 ft)	1975	Road	Uruguay and Argentina
Pamban Bridge (Road)	2345 m (7694 ft)	115 m (377 ft)	1988	Road	India
Marabá Mixed Bridge	2340 m (7680 ft)	?	1984		Brazil
Juan Pablo II Bridge	2310 m (7580 ft)	?	1974	Road	Chile
Evergreen Point Floating Bridge	2310 m (7580 ft)	?	1963	Highway	United States
Rho Viaduct[50]	2300 m (7500 ft)	?	2007		Italy
Fuller Warren Bridge	2286 m (7500 ft)	76 m (249 ft)	2002	Highway	United States
Ambassador Bridge	2283 m (7490 ft)	564 m (1850 ft)	1929	Highway	Canada and United States
25 de Abril Bridge	2278 m (7474 ft)	1013 m (3323 ft)	1966	Road & rail	Portugal
Mahanadi River Bridge	2258 m (7408 ft)	?	?	Highway & Railroad	India
Maurício Joppert Bridge	2250 m (7380 ft)	112 m (367 ft)	1964		Brazil
Teodoro Moscoso Bridge	2250 m (7380 ft)	?	1993	Highway	Puerto Rico
Tay Road Bridge	2250 m (7380 ft)	?	1966	Road & pedestrian	United Kingdom
Beška Bridge[51]	2250 m (7380 ft)	210 m (690 ft)	1975	Highway	Serbia
Jamestown-Verrazano Bridge	2240 m (7350 ft)	183 m (600 ft)	1992	Highway	United States
Yen Lenh Bridge	2230 m (7320 ft)		2004	Road	Vietnam

McKees Rocks Bridge	2225 m (7300 ft)	229 m (751 ft)	1931	Highway	United States
Anping Bridge[52]	2223 m (7293 ft)	?	1151	Pedestrian	People's Republic of China
Humber Bridge	2220 m (7280 ft)	1410 m (4630 ft)	1981	Road, pedestrian & cyclist	United Kingdom
Novo Oriente Bridge	2200 m (7200 ft)	50 m (160 ft)	1990		Brazil
Tsing Ma Bridge	2200 m (7200 ft)	1377 m (4518 ft)	1997	Road & rail	Hong Kong
San Juanico Bridge	2200 m (7200 ft)	1377 m (4518 ft)	1979	Road & pedestrian	Philippines
Abraham Lincoln Memorial Bridge	2170 m (7120 ft)	189 m (620 ft)	1987	Road	United States
Llacolen Bridge[53]	2157 m (7077 ft)	?	2000		Chile
Novosibirsk Metro Bridge	2145 m (7037 ft)	?	1986	Rail (Metro)	Russia
Pont de Normandie	2141 m (7024 ft)	856 m (2808 ft)	1995	Road	France
Igelsta Bridge	2140 m (7020 ft)	100 m (330 ft)	1995	Rail	Sweden
Viadotto San Floriano[54]	3567 m (11703 ft)	?	1990		Italy
Surgut Bridge	2110 m (6920 ft)	408 m (1339 ft)	2000	Road	Russia
Phu My Bridge	2230 m (7320 ft)	380 m (1250 ft)	2009	Road	Vietnam
Des Plaines River Valley Bridge	2100 m (6900 ft)	?	2007	Highway	United States
Port Mann Bridge	2093 m (6867 ft)	366 m (1201 ft)	1964	Road	Canada
Manhattan Bridge	2089 m (6854 ft)	448 m (1470 ft)	1909	Highway & Railroad	United States
Pamban Bridge (Rail)	2065 m (6775 ft)	115 m (377 ft)	1914	Rail	India
Poughkeepsie Bridge	2064 m (6772 ft)	160 m (520 ft)	1889	Pedestrian	United States
Sharavathi Bridge (Honnavar Bridge)[55]	2060 m (6760 ft)	33 m (108 ft)	1994	Rail	India
Asparuhov Most	2050 m (6730 ft)	160 m (520 ft)	1976	Road	Bulgaria
Verrazano-Narrows Bridge	2034 m (6673 ft)	1298 m (4259 ft)	1964	Highway	United States
Lacey V. Murrow Memorial Bridge	2019 m (6624 ft)	?	1993	Highway	United States
Delaware River-Turnpike Toll Bridge	2003 m (6572 ft)	208 m (682 ft)	1956	Highway	United States

Under construction

This transport-related list is incomplete; you can help by expanding it [56].

Name	Length kilometres (miles)	Span metres (feet)	Year	Traffic	Country
Qatar-Bahrain Causeway[57]	40 km (25 mi)		2013		▬ Qatar and ▬ Bahrain
Louisiana Highway 1 bridge[58]	29 km (18 mi)[21]		2011	Highway	▬ United States
Bridge of the Horns	29 km (18 mi)		2020	Road	▬ Yemen and ▬ Djibouti
Jaber Bridge	24.1 km (15.0 mi)	300 m (980 ft)	2015	Road	▬ Madinat Al-Hareer, Kuwait
Penang Second Bridge	23.4 km (14.5 mi)	250 m (820 ft)	2012	Road	▬ Malaysia
Hong Kong-Zhuhai-Macau Bridge (main bridge)	22.8 km (14.2 mi)	460 m (1510 ft)	2016	Highway	▬ People's Republic of China
Mumbai Trans Harbour Link	22 km (14 mi)		2014	Road and Metro	▬ India
Hami Grade Separation Bridge *Lanzhou–Urumqi High-Speed Railway*	19.3 km (12.0 mi)[59]		2013	High-speed rail	▬ People's Republic of China
Chennai Port – Maduravoyal Expressway	19 km (12 mi)	1932 m (6339 ft)	2009, completed in 2013	Road	▬ India
Third Orinoco crossing (Mercosur Bridge)[60] [61]	11.1 km (6.9 mi)		2011		▬ Venezuela
Podilsko-Voskresensky Bridge[62]	7.04 km (4.37 mi)		2012	Road & Metro	▬ Ukraine
Ershilipu Bridge *Lanzhou–Urumqi High-Speed Railway*	7.0926 km (4.4071 mi)[63]		2013	High-speed rail	▬ People's Republic of China
Jajmau Bypass	5.0 km (3.1 mi)		2013	Road	▬ India
Bogibeel Bridge[64]	4.3 km (2.7 mi)		2013	Road & Rail	▬ India
Macleay River Bridge[65]	3.2 km (2.0 mi)		2013	Road	▬ Australia
Forth Replacement Crossing	2.7 km (1.7 mi)[66]		2016	Road	▬ United Kingdom
Persian Gulf Causeway *Khaleej-e Fars (Laft-Pohl) Bridge*	2.2 km (1.4 mi)[67]		2014	High-speed railroad & pipeline	Qeshm Island, ▬ Iran

See also

- List of longest masonry arch bridge spans
- List of longest suspension bridge spans
- List of longest cantilever bridges
- List of longest continuous truss bridge spans
- List of largest cable-stayed bridges
- List of bridges
- List of tunnels by length

References

[1] Guinness World Record, Longest bridge (http://www.guinnessworldrecords.com/Search/Details/Longestbridge/49848.htm), Guinness World Records. Last accessed July 2011.

[2] http://wenku.baidu.com/view/42959918964bcf84b9d57bdd.html

[3] 1228 (http://xian.qq.com/a/20090506/000004.htm)

[4] Guinness World Record, Longest bridge over water (aggregate length) (http://www.guinnessworldrecords.com/Search/Details/Longestbridge-over-water-(aggregate-length)/76012.htm), Guinness World Records. Last accessed July 2011.

[5] Qingdao Haiwan Bridge

[6] 26.75+5.85+0.9 Bridge engineering in China (http://civil.fzu.edu.cn/BridgeCourseAttachment/200905/bridge engineering in China(4).pdf)

[7] Guinness World Record, Longest bridge over water (continuous length) (http://www.guinnessworldrecords.com/Search/Details/Longestbridge-over-water-(continuous)/76642.htm), Guinness World Records. Last accessed July 2011.

[8] (http://att.newsmth.net/att.php?p.443.327603.703.jpg)

[9] (http://www.stbridge.com.cn/intro/default.asp)

[10] "6 Zusammenfassung, Schlussfolgerungen und Empfehlungen" (http://www.tu-dresden.de/vkiva/hlb/fachtagung_tr/trt4/12_trt4_vortrag12.pdf) (PDF). . Retrieved 2010-09-06.

[11] (http://www.china-rail.com.cn/yjhq/2007/200706/2007-06-04/20070604102954_56118.html)

[12] http://www.ghumakkar.com/2009/08/07/delhi-kanpur-lucknow-gorakhpur-road-trip/

[13] Sara United Advertising, Alkhobar, Saudi Arabia. "::: King Fahd Causeway Authority ::: Saudi Arabia, Bahrain" (http://www.kfca.com.sa/about_tech.htm). Kfca.com.sa. . Retrieved 2010-09-06.

[14] (http://att.newsmth.net/att.php?p.443.327603.869300.jpg)

[15] C215 Viaduct (http://en.structurae.de/structures/data/index.cfm?ID=s0004210) at *Structurae*

[16] "Incheon-bridge.com - de beste bron van informatie over incheon bridge" (http://eng.incheon-bridge.com/02_introduction/summary/main02.htm). Eng.incheon-bridge.com. . Retrieved 2010-09-06.

[17] http://www.ng.ru/regions/2002-09-12/4_kazan.html

[18] "ВОЛГОМОСТ —Ульяновский мост будет сдан в следующем году" (http://www.volgomost.ru/pressroom/release/104.html). Volgomost.ru. . Retrieved 2010-09-06.

[19] New bridge in Saratov (russian) (http://www.ng.ru/events/2000-11-22/2_newbridge.html)

[20] Nantong Yangshan Yellow-Sea Bridge (http://politics.people.com.cn/GB/14562/7525545.html)

[21] 90%"" (http://news.xinhuanet.com/politics/2008-04/22/content_8025633.htm)

[22] Qinghai-Tibet Railway (http://www.tibetinfor.com/english/zt/040719_qztl/..\040719_qztl/200402004727155221.htm)

[23] BRISA (http://www.brisa.pt/Brisa/vPT/Centro+de+Imprensa/Noticias/RepositÃ³rio/PonteLezÃria.htm)

[24] The total length of all raised spans is 1.9 miles (3.1 km), but the shipping channel span is 750 feet (229 m) Bay Area Toll Authority - Bridge Facts - San Mateo-Hayward Bridge (http://bata.mtc.ca.gov/bridges/sm-hayward.htm)

[25] Chang Dong Yellow River Bridge (http://en.structurae.de/structures/data/index.cfm?ID=s0006686) at *Structurae*

[26] Xiasha Bridge (http://en.structurae.de/structures/data/index.cfm?ID=s0006002) at *Structurae*

[27] Jiujiang Yangtze River Bridge (http://en.structurae.de/structures/data/index.cfm?ID=s0005673) at *Structurae*

[28] Seohae Bridge (http://en.structurae.de/structures/data/index.cfm?ID=s0000614) at *Structurae*

[29] Nationalbridges.com National Bridge Inventory Bridges (http://nationalbridges.com/nbi_record.php?StateCode=37&struct=000000001870015)

[30] North Luzon Expressway#List of exits

[31] Nationalbridges.com National Bridge Inventory Bridges (http://nationalbridges.com/nbi_record.php?StateCode=37&struct=000000001770007)

[32] Jingzhou Yangtze River Bridge (http://en.structurae.de/structures/data/index.cfm?ID=s0001855) at *Structurae*

[33] "30th Yangtze River Bridge Ends Need for Local Ferry" (http://english.peopledaily.com.cn/200210/02/eng20021002_104313.shtml). English.peopledaily.com.cn. 2002-10-02. . Retrieved 2010-09-06.

[34] Elbmarsch-Hochbrücke (1976) (http://en.structurae.de/structures/data/index.cfm?ID=s0004993) at *Structurae*

[35] National Bridge Inventory Bridges (http://nationalbridges.com/nbi_record.php?StateCode=37&
struct=000000000550011Nationalbridges.com)

[36] "Queen Isabella Causeway - Wikipedia, the free encyclopedia" (http://en.wikipedia.org/wiki/Queen_Isabella_Causeway).
En.wikipedia.org. . Retrieved 2010-09-06.

[37] "Mozambique News Agency - AIM Reports" (http://www.poptel.org.uk/mozambique-news/newsletter/aim281.html#story9).
Poptel.org.uk. 2004-08-12. . Retrieved 2010-09-06.

[38] (http://www.rfi.it/files/varie/The new high speed TURIN - MILAN line.pdf)

[39] Target Bridge Window For Japan (http://www.daido-it.ac.jp/~doboku/miki/inxjb3.html)

[40] Fadalto Viaduct (http://en.structurae.de/structures/data/index.cfm?ID=s0008594) at *Structurae*

[41] (http://www.brdi.com.cn/english/bridges/arch_changsha.htm), (http://english.people.com.cn/200405/26/eng20040526_144403.
html)

[42] "Longest Railway Bridge in India - Nehru Setu on River Sone" (http://www.thecolorsofindia.com/interesting-facts/infrastructure/
longest-railway-bridge-in-india.html). Thecolorsofindia.com. . Retrieved 2010-09-06.

[43] C310 Viaduct (http://en.structurae.de/structures/data/index.cfm?ID=s0004209) at *Structurae*

[44] Re Island Bridge (http://en.structurae.de/structures/data/index.cfm?ID=s0002369) at *Structurae*

[45] Oleron Bridge (http://en.structurae.de/structures/data/index.cfm?ID=s0000150) at *Structurae*

[46] Second Quiantang River Bridge (http://en.structurae.de/structures/data/index.cfm?ID=s0005666) at *Structurae*

[47] Banghwa Bridge (http://www.poonglim.co.kr/eng/busi_area/01load/tunnel_01.htm)

[48] Kremsbrücke Pressingberg (1980) (http://en.structurae.de/structures/data/index.cfm?ID=s0008589) at *Structurae*

[49] Drežnik Viadukt (http://en.structurae.de/structures/data/index.cfm?ID=s0010103) at *Structurae*

[50] Rho Viaduct (http://en.structurae.de/structures/data/index.cfm?ID=s0019595) at *Structurae*

[51] Plovput |> MOST : Beška (http://www.plovput.rs/plovni_put_mostovi_full_1232.htm)

[52] Anping Bridge (http://en.structurae.de/structures/data/index.cfm?ID=s0005531) at *Structurae*

[53] Visit to the Legendary Bío-Bío (http://www.welcomechile.com/concepcion/biobio-river.html)

[54] Viadotto San Floriano (http://en.structurae.de/structures/data/index.cfm?ID=s0008591) at *Structurae*

[55] "Sharavati Bridge" (http://www.afcons.com/Bridges/Brides_complited/KonkanRailway.html). Afcons Infrastructure Limited. .
Retrieved 2012-01-07.

[56] http://en.wikipedia.org/wiki/List_of_longest_bridges_in_the_world

[57] "Quatar [sic] et Bahreïn reliés par un pont de 40 km en 2013" (http://metropolegeneve.blog.tdg.ch/archive/2008/05/07/
quatar-et-bahrein-relie-par-un-pont-de-40-km-2013.html), *La Tribune de Geneve*, May 5, 2008

[58] LA1 Project (http://www.la1project.com/)

[59] http://www.tianshannet.com.cn/news/content/2010-12/12/content_5436119.htm

[60] - Tercer puente sobre el Orinoco registra 24 por ciento de avance (http://www.radiomundial.com.ve/yvke/noticia.php?6164)

[61] - Tercer puente sobre el río Orinoco se llamará Mercosur (http://www.mci.gob.ve/noticias_-_prensa/28/10820/tercer_puente_sobre.
html)

[62] Podilskyi Metro Bridge (http://en.structurae.de/structures/data/index.cfm?ID=s0025419) at *Structurae*, Report on the official hearing
on construction progress (http://www.archunion.com.ua/sovet-2006/gradsovet_06_05_24.shtml)

[63] http://news.xinmin.cn/rollnews/2010/12/15/8310728.html

[64] (http://www.assamtribune.com/scripts/detailsnew.asp?id=may0811/at092)

[65] Kempsey Bypass Project Overview (http://www.rta.nsw.gov.au/roadprojects/projects/pac_hwy/port_macquarie_coffs_harbour/
kempsey_bypass/project_overview.html)

[66] http://www.roadtraffic-technology.com/projects/forth-crossing/

[67] http://www.presstv.ir/detail/169858.html

Hrvatske_ceste

Type	State-owned limited company
Industry	Road transport
Founded	2001
Headquarters	Zagreb, Croatia
Key people	Jakov Krešić
Website	[1]

Hrvatske ceste (lit. *Croatian roads*) is a Croatian state-owned company pursuant to provisions of the Croatian Public Roads Act (Croatian: *Zakon o javnim cestama* enacted by the Parliament of the Republic of Croatia.[2] Tasks of the company are defined by Public Roads Act and its Founding Declaration, and the principal task of the company is management, construction and maintenance of public roads. In practice, Hrvatske ceste are responsible for the State roads in Croatia (designated with *D*), the county (*Ž*) and local (*L*) roads are managed by county authorities, while the motorways (*A*) are managed by Croatian Motorways Ltd and other concessionaires.

The company is currently administered by a three-person managing board consisting of Jakov Krešić (chairman and Josip Škorić and five-member supervisory board.[3] [4]

The company was first established on April 6, 2001, under the law promulgated on on April 5, 2001,[2] with the share capital of the company worth 128,898,200.00 Croatian kuna.

Hrvatske ceste are organized in six business sectors: Studies and design, Construction, Maintenance, Procurement, Financial and business operations and Legal, personnel and general sectors.[5]

All profits generated by Hrvatske ceste are used for construction and maintenance of the roads the company manages.

See also

- State roads in Croatia

References

[1] http://www.hrvatske-ceste.hr/
[2] "Zakon o javnim cestama - Public Roads Act" (http://narodne-novine.nn.hr/clanci/sluzbeni/2004_12_180_3130.html) (in Croatian). *Narodne novine*. December 14, 2004. .
[3] "Hrvatske ceste - management" (http://www.hrvatske-ceste.hr/Uprava-drustva.htm) (in Croatian). *Hrvatske ceste*. June 27, 2010. .
[4] "Hrvatske ceste - supervisory board" (http://www.hrvatske-ceste.hr/Nadzorni-odbor.htm) (in Croatian). *Hrvatske ceste*. June 27, 2010. .
[5] "Hrvatske ceste - structure" (http://www.hrvatske-ceste.hr/Ustroj.htm) (in Croatian). *Hrvatske ceste*. June 27, 2010. .

External links

- Official website (http://http://www.hrvatske-ceste.hr/)

Article Sources and Contributors

Dobra_Bridge_(A6) *Source*: http://en.wikipedia.org/w/index.php?title=Dobra_Bridge_%28A6%29 *Contributors*: Tomobe03

A6_(Croatia) *Source*: http://en.wikipedia.org/w/index.php?title=A6_%28Croatia%29 *Contributors*: Admiral Norton, Brutaldeluxe, Diannaa, Dr. Blofeld, Giraffedata, GregorB, Imzadi1979, Joy, LilHelpa, Maartenvdbent, Maestral, MarkBA, Niagara, Nono64, Rjwilmsi, Roberta F., Tabletop, Tagishsimon, Thomprod, Tomobe03, WOSlinker, Zscout370, 6 anonymous edits

Gorski_kotar *Source*: http://en.wikipedia.org/w/index.php?title=Gorski_kotar *Contributors*: 1331331331, Antije, Cmdrjameson, CrniBombarder!!!, Croat Canuck, Croatiaundiscovered, DerBorg, EugeneZelenko, Evdin, Gaius Cornelius, GregorB, Guy0307, Izvora, Jesuislafete, Joekoz451, Jonel, Jorunn, Joy, Kubura, Kwamikagami, Litany, Luka Jačov, Luna Santin, Mihovil, Nick Number, PANONIAN, Piastu, Simetrical, SkerHawx, TimBentley, Tomobe03, Woohookitty, Xpatrik, Ziga, Zscout370, Zupanja99, 24 anonymous edits

Dobra_(river) *Source*: http://en.wikipedia.org/w/index.php?title=Dobra_%28river%29 *Contributors*: Brajski paur, D6, Darwinek, Discospinster, Joy, Kebeta, Markussep, Sylfred1977, TexasAndroid, 2 anonymous edits

Ticket_system *Source*: http://en.wikipedia.org/w/index.php?title=Ticket_system *Contributors*: Admiral Norton, AdriansyahYS, Bennyzhong, Coolcaesar, Dough4872, Hmains, MC10, OldsVistaCruiser, Pascal.Tesson, Rich Farmbrough, Rwboa22, Sonjaaa, Svick, Wongm, Wxstorm, 10 anonymous edits

Autocesta_Rijeka_–_Zagreb *Source*: http://en.wikipedia.org/w/index.php?title=Autocesta_Rijeka_%E2%80%93_Zagreb *Contributors*: Timbouctou, Tomobe03

Adriatic_Sea *Source*: http://en.wikipedia.org/w/index.php?title=Adriatic_Sea *Contributors*: 12dstring, AJR, Admiral Norton, Adriatikus, AjaxSmack, Akanemoto, Alansohn, Alban.bytyci, Alias Flood, Alphax, Amarkov, Andre Engels, Arthurian Legend, Asim Led, August Dominus, Avala, Avatarion, Banjee ca, Bardsandwarriors, Bazonka, Beginning, Bkonrad, Bludyta, Bogdangiusca, Bomac, Bosniak, Bratislav, Brighterorange, Bryan Derksen, CommonsDelinker, Conversion script, Cowpilot, Croacting77, Cyrus the virus grissimo, DIREKTOR, Daesitiates, Dalmatine, DandyDan2007, DarkoS, Dbachmann, Dbo789, Deflective, Den fjättrade ankan, Dietary Fiber, Dinesh smita, Dori, Downaware, Dromadar, DuncanHill, Dutchtower, Edolen1, Edwinstearns, El C, Eleassar, Emijrp, Esperant, Estudiarme, Eu.stefan, Evercat, Fastifex, Flibjib8, Freddyballo, Freshprince5252, Frokor, Gene Nygaard, Giespe, Gilliam, Gkmx, Glenn, Gonzonoir, GospodarSvjetlosti, Graham87, GregorB, Grymnir, Harioris, HarisM, Herbythyme, Hmains, Husky, IanDavies, Idaltu, Igor, Ilva, Jadvinia, Jagun, Jajogluck, Jakiša Tomić, Jamie C, Janke, Jauhienij, Jazavac, Jdpockar, Jimp, Joy, Juliancolton, Jumbuck, Juzeris, Kebeta, Khfinch, Khoikhoi, Klemen Kocjancic, Kluka, Kokox, Kungfuadam, Kwamikagami, La Pianista, Laser brain, Lazerproofsandals, Lightmouse, Lothar Klaic, Lucinos, Luna Santin, Ma7trix7, MaNeMeBasat, Maestral, Mani1, Marek69, Marquez, Mattis, Mattius Dearius, Mav, Maxim Razin, Mazbln, Mdeegan, Menchi, Mestric, Miaow Miaow, Michellecrisp, Mike Rosoft, Modeha, Mschlindwein, Muleni, My little halo, Naddy, Natalino7, NawlinWiki, Neddyseagoon, NeroN BG, Newkai, Nikola Smolenski, Ninjalemming, Noclador, Nocturnal Wanderer, NormanEinstein, Northamerica1000, Notscott, Npeters22, Nsaa, Nxghzt, Odie5533, Oldlaptop321, Originalwana, Orsat1, PANONIAN, Para, Paul Drye, Pax:Vobiscum, Paytim, Peaches111, Peter Greenwell, Peterlin, Piratejesus, Poromiami, Ppntori, Rasho, Reeve, Rejectwater, Rezwalker, Rjwilmsi, Rokpok, Ronhjones, Rosiestep, Rsnetto74, RunningMan, Sammy3434, Seba, Sebleblanc, Selma Kaufmann, Shephia, Shikuesi3, Shyam, Sideshow Bob, Simone, Slothropslothrop, Sortior, South Bay, SpookyMulder, StassiNet, Staxringold, Steven Weston, Sulmues, Taulant23, The Epopt, Thomas Larsen, Thricecube, Timrollpickering, Tolo, Tomobe03, Tygrrr, Uirauna, Urmas, UtherSRG, VVVladimir, Vanished user 03, Vicki Rosenzweig, Vsmith, W Auckland, Wetman, Wheneverwhenever, Wikigregor, Wmahan, Woloflover, Woohookitty, WordsExpert, XJamRastafire, Zanhe, Zenanarh, Zocky, Zoran Knez, Александър, 254 anonymous edits

List_of_longest_bridges_in_the_world *Source*: http://en.wikipedia.org/w/index.php?title=List_of_longest_bridges_in_the_world *Contributors*: 000peter, 84user, Aaron charles, Adam Bishop, Adam78, Adityavijay09, Advanstra, Ae-a, Aiman abmajid, Airplaneman, Alanmak, Alchemistmatt, Aliasd, Alokprasad, Amckern, AntoniusJ, Ardric47, Aridd, Aries1975, Ashish itct, AtilimGunesBaydin, Auric, Australiss, Autonic, Aylahs, BIL, Beerandt, Bendono, Betenner, Bhonsley, Bilsonius, Bine Mai, Bishupriyaparam, BlckKnght, Blue Elf, Brunton, Bull22289, BuzzfledderJohn, ByeByeBaby, Cabazap, Cacophony, Capricorn42, CeeGee, Chandan Guha, Channelologist, Charles Matthews, CharlotteWebb, Chaser, Chill doubt, ChrisRuvolo, Cmghim925, CommonsDelinker, Coopman86, Courcelles, Cprice648, D, DJ Clayworth, Dale Arnett, Danielcohadier, Darkdroom, Dars-dm, Ddama, Ddstretch, Dejiolowe, Derekjoe, Devoxo, Djzare, DI2836, Dontworry, Doradus, Douglas the Comeback Kid, Drenaline, Dunncon13, Dyaimz, Dynamization, EaglesFanInTampa, Edolen1, Effer, EllisB123, Emarsee, Encyclopedia25, Engi08, Eraserhead1, Eugene van der Pijll, Falcon8765, Filipdr, Gabbe, Generalhuo, Ghirlandajo, Gleppe, Gmorris3589, Gooday.1, Gosox, Grayshi, Green Cardamom, Grenavitar, Greyhood, Gökhan, Hankhuck, Hawarnekar, Hhgygy, Hqb, Imachitch, In fact, Instantnood, Ioldanach, Irkthesam, Iva T, J intela, Jaiprakashsingh, Jaksmata, JalF, JamesBWatson, Jason041, Jeffreyarcand, JenniferHeartsU, Jeronim, Jerryseinfeld, Jicosa, JimVC3, Jklamo, Joeylawn, JohnCD, JonRoma, Jpatokal, Jpbowen, Jpgordon, KelleyCook, Keraunos, KevinGarnerNC, Kilo-Lima, KimvdLinde, Klaus with K, Kmsiever, Koavf, Kotukunui, Kshitij1508, Kvdh, Kzaral, LOL, LUxlii, Le Chiffre17, LeadSongDog, Lotheric, Lozen130, Lsxxsc, Luis wiki, Luka Jačov, MTC, Mad Greg, MakeChooChooGoNow, Makedonas, Marcopolo112233, Marut28, Maurice Carbonaro, Megaproj, Meheureux, MickMacNee, Midad Arif, Miraceti, Mistakefinder, Montrealais, Mr.Clown, Mrcool1122, Mulad, Multivariable, Mwanner, Mygonad, Mythbuster2010, NE2, Najro, Nat Krause, Nbumbic, Newmanbe, Nibblesnbits, Night Tracks, Nima Farid, Nirvana888, Nottheking, Orudge, Overandaway, Pantjz, Patriarca12, Patrick, PeterFisk, Pieleric, Pinas Central, Plastikspork, Polylepsis, Pqrshanth, Prenn, Presidentman, Priyankoo, Prvc, Pumpkin414, Python eggs, Qertis, Qqqqqq, QuadrivialMind, Quantpole, Radustoicescu, Ragib, Rameshnz, Rashat999, Recurring dreams, Redlioness, Regregex, Rehrenberg, Reign of Toads, Rich Farmbrough, Richard David Ramsey, Robert Merkel, Rockingthiru, Rohith goura, Ronita saha, Rontombontom, Roulex 45, Rye1967, SPUI, Sam, San Sanitsch, Sanko, SchmuckyTheCat, Sciurinæ, ScottDavis, Sellyme, Sentausa, Seth harsch, Setti, Seuraza, Sgt Simpson, Shylocxs, Simi8017, Sinolonghai, Sirjasono, Skaperat, Skew-t, Sobreira, Softjuice, Stefan Jansen, Stephen Morley, Sungminkwon, Swilburn, TFOWR, TYW, Tabletop, Taro 80, TarzanASG, Tawker, Thecheesykid, Thryduulf, Timsdad, Tnds, Tomdoan, Tommyt, Tomobe03, Tphcm, Trainrider002, Trikur, Triona, Tuduser, Twas Now, Unschool, Unstudmaddu, Urhixidur, Vegaswikian, VerruckteDan, Vijethnbharadwaj, Vir iv, Wacko, Wcivils, Weirdy, West.andrew.g, WikiDao, Wikid77, Wikikeef, Wikiolap, Winhunter, Wolfrock, Woohookitty, XLerate, YADJML, Yossiea, Zarcadia, Zimbabweed, Zinc2005, Zippanova, Zocky, Zollerriia, جماعدلي.24, 648 anonymous edits

Hrvatske_ceste *Source*: http://en.wikipedia.org/w/index.php?title=Hrvatske_ceste *Contributors*: GregorB, Joy, Tim!, Tomobe03, 2 anonymous edits

Image Sources, Licenses and Contributors

File:M Dobra A6 0509.jpg *Source*: http://en.wikipedia.org/w/index.php?title=File:M_Dobra_A6_0509.jpg *License*: unknown *Contributors*: User:Roberta F.

File:Autocesta A6.svg *Source*: http://en.wikipedia.org/w/index.php?title=File:Autocesta_A6.svg *License*: unknown *Contributors*: User:Tomobe03

File:Croatia Autocesta A6.svg *Source*: http://en.wikipedia.org/w/index.php?title=File:Croatia_Autocesta_A6.svg *License*: unknown *Contributors*: User:Jeremiah21, User:Minestrone

File:Chs2 Greend40.png *Source*: http://en.wikipedia.org/w/index.php?title=File:Chs2_Greend40.png *License*: unknown *Contributors*: BrokenSegue, Klin, Paradoctor, Tegu

File:Color icon Cornflower blue.svg *Source*: http://en.wikipedia.org/w/index.php?title=File:Color_icon_Cornflower_blue.svg *License*: unknown *Contributors*: User:FedotPraslov

Image:E65-HR.svg *Source*: http://en.wikipedia.org/w/index.php?title=File:E65-HR.svg *License*: unknown *Contributors*: User:Rl91

File:Autocesta A1.svg *Source*: http://en.wikipedia.org/w/index.php?title=File:Autocesta_A1.svg *License*: unknown *Contributors*: User:Tomobe03

File:Državna cesta D3.svg *Source*: http://en.wikipedia.org/w/index.php?title=File:Državna_cesta_D3.svg *License*: unknown *Contributors*: User:Jure Grm

File:Državna cesta D501.svg *Source*: http://en.wikipedia.org/w/index.php?title=File:Državna_cesta_D501.svg *License*: unknown *Contributors*: User:Jure Grm

File:Državna cesta D40.svg *Source*: http://en.wikipedia.org/w/index.php?title=File:Državna_cesta_D40.svg *License*: unknown *Contributors*: Tomobe03

File:Autocesta A7.svg *Source*: http://en.wikipedia.org/w/index.php?title=File:Autocesta_A7.svg *License*: unknown *Contributors*: User:Tomobe03

File:Bosiljevo Croatia A1-A6.jpg *Source*: http://en.wikipedia.org/w/index.php?title=File:Bosiljevo_Croatia_A1-A6.jpg *License*: unknown *Contributors*: User:Ex13

File:Cvor Orehovica Rijeka 060708.jpg *Source*: http://en.wikipedia.org/w/index.php?title=File:Cvor_Orehovica_Rijeka_060708.jpg *License*: unknown *Contributors*: User:Roberta F.

File:A6 upozorenja 0509.jpg *Source*: http://en.wikipedia.org/w/index.php?title=File:A6_upozorenja_0509.jpg *License*: unknown *Contributors*: User:Roberta F.

File:I Ostrovica A6 0509.jpg *Source*: http://en.wikipedia.org/w/index.php?title=File:I_Ostrovica_A6_0509.jpg *License*: unknown *Contributors*: User:Roberta F.

File:Bajer lake, Gorsko Kotar, Croatia.JPG *Source*: http://en.wikipedia.org/w/index.php?title=File:Bajer_lake,_Gorsko_Kotar,_Croatia.JPG *License*: unknown *Contributors*: User:Modzzak

File:T Podvugles A6 0509.jpg *Source*: http://en.wikipedia.org/w/index.php?title=File:T_Podvugles_A6_0509.jpg *License*: unknown *Contributors*: User:Roberta F.

File:T Tuhobic A6 0509.jpg *Source*: http://en.wikipedia.org/w/index.php?title=File:T_Tuhobic_A6_0509.jpg *License*: unknown *Contributors*: User:Roberta F.

Image:A6 AADT 2009.gif *Source*: http://en.wikipedia.org/w/index.php?title=File:A6_AADT_2009.gif *License*: unknown *Contributors*: User:Tomobe03

File:A6 12 o Ravna Gora 130109.jpg *Source*: http://en.wikipedia.org/w/index.php?title=File:A6_12_o_Ravna_Gora_130109.jpg *License*: unknown *Contributors*: User:Roberta F.

File:E65-HR.svg *Source*: http://en.wikipedia.org/w/index.php?title=File:E65-HR.svg *License*: unknown *Contributors*: User:Rl91

File:Državna cesta D42.svg *Source*: http://en.wikipedia.org/w/index.php?title=File:Državna_cesta_D42.svg *License*: unknown *Contributors*: User:Jure Grm

Image:Zeichen 314.svg *Source*: http://en.wikipedia.org/w/index.php?title=File:Zeichen_314.svg *License*: unknown *Contributors*: Andreas 06, Carl Auer, Cfaerber, Cäsium137, Juiced lemon, MB-one, RTCNCA, Rfc1394, Túrelio

File:Z5034-HR.svg *Source*: http://en.wikipedia.org/w/index.php?title=File:Z5034-HR.svg *License*: unknown *Contributors*: User:Rl91

File:Z5068-HR.svg *Source*: http://en.wikipedia.org/w/index.php?title=File:Z5068-HR.svg *License*: unknown *Contributors*: User:Rl91

Image:Zeichen 391.svg *Source*: http://en.wikipedia.org/w/index.php?title=File:Zeichen_391.svg *License*: unknown *Contributors*: Andreas 06, Augiasstallputzer, Burts, Cfaerber, Cäsium137, Umherirrender

File:E61-HR.svg *Source*: http://en.wikipedia.org/w/index.php?title=File:E61-HR.svg *License*: unknown *Contributors*: User:Rl91

File:Z5054-HR.svg *Source*: http://en.wikipedia.org/w/index.php?title=File:Z5054-HR.svg *License*: unknown *Contributors*: User:Rl91

File:LokvarskoJezero1.jpg *Source*: http://en.wikipedia.org/w/index.php?title=File:LokvarskoJezero1.jpg *License*: unknown *Contributors*: User:Sl-Ziga

Image:Paturnpiketicket.jpg *Source*: http://en.wikipedia.org/w/index.php?title=File:Paturnpiketicket.jpg *License*: unknown *Contributors*: Clinton N. Godlesky

File:Državna cesta D102.svg *Source*: http://en.wikipedia.org/w/index.php?title=File:Državna_cesta_D102.svg *License*: unknown *Contributors*: User:Jure Grm

File:Loudspeaker.svg *Source*: http://en.wikipedia.org/w/index.php?title=File:Loudspeaker.svg *License*: unknown *Contributors*: Bayo, Gmaxwell, Husky, Iamunknown, Mirithing, Myself488, Nethac DIU, Omegatron, Rocket000, The Evil IP address, Wouterhagens, 18 anonymous edits

File:Zatoka Kotorska.jpg *Source*: http://en.wikipedia.org/w/index.php?title=File:Zatoka_Kotorska.jpg *License*: unknown *Contributors*: User:Janusz Recław

File:Adriatic_Sea_islands.jpg *Source*: http://en.wikipedia.org/w/index.php?title=File:Adriatic_Sea_islands.jpg *License*: unknown *Contributors*: Bukk, Cro-Cop2, EPO, Veliki Kategorizator

File:Magnify-clip.png *Source*: http://en.wikipedia.org/w/index.php?title=File:Magnify-clip.png *License*: unknown *Contributors*: User:Erasoft24

File:Adriatic jadran.png *Source*: http://en.wikipedia.org/w/index.php?title=File:Adriatic_jadran.png *License*: unknown *Contributors*: Bratislav Tabaš

File:Vrulja kod Omiša.jpg *Source*: http://en.wikipedia.org/w/index.php?title=File:Vrulja_kod_Omiša.jpg *License*: unknown *Contributors*: User:Ante Perkovic

File:Bari02.jpg *Source*: http://en.wikipedia.org/w/index.php?title=File:Bari02.jpg *License*: unknown *Contributors*: FlickreviewR, Leoboudv, MECU, Mac9, Shizhao

File:Venezia veduta aerea.jpg *Source*: http://en.wikipedia.org/w/index.php?title=File:Venezia_veduta_aerea.jpg *License*: unknown *Contributors*: User:Oliver-Bonjoch

File:Trieste-IMG 3064.JPG *Source*: http://en.wikipedia.org/w/index.php?title=File:Trieste-IMG_3064.JPG *License*: unknown *Contributors*: User:Rinina25, User:Twice25

File:City-of-Split.jpg *Source*: http://en.wikipedia.org/w/index.php?title=File:City-of-Split.jpg *License*: unknown *Contributors*: User:Lpuljak

File:MOSE Project Venice from the air.jpg *Source*: http://en.wikipedia.org/w/index.php?title=File:MOSE_Project_Venice_from_the_air.jpg *License*: unknown *Contributors*: Chris 73

File:Adriatic Plate.jpg *Source*: http://en.wikipedia.org/w/index.php?title=File:Adriatic_Plate.jpg *License*: unknown *Contributors*: User:Sting, User:קיץ

File:Spring Runoff in the Adriatic Sea.jpg *Source*: http://en.wikipedia.org/w/index.php?title=File:Spring_Runoff_in_the_Adriatic_Sea.jpg *License*: unknown *Contributors*: Jeff Schmaltz

File:Tremiti 00.jpg *Source*: http://en.wikipedia.org/w/index.php?title=File:Tremiti_00.jpg *License*: unknown *Contributors*: User:Raboe001

File:Kornati islands.jpg *Source*: http://en.wikipedia.org/w/index.php?title=File:Kornati_islands.jpg *License*: unknown *Contributors*: user:donarreiskoffer

File:Pula Arena aerial 1.jpg *Source*: http://en.wikipedia.org/w/index.php?title=File:Pula_Arena_aerial_1.jpg *License*: unknown *Contributors*: Boris Licina

File:Venezia Riva degli Schiavoni 002.JPG *Source*: http://en.wikipedia.org/w/index.php?title=File:Venezia_Riva_degli_Schiavoni_002.JPG *License*: unknown *Contributors*: User:Janericloebe

File:Battle of Lissa.jpg *Source*: http://en.wikipedia.org/w/index.php?title=File:Battle_of_Lissa.jpg *License*: unknown *Contributors*: Engraving by Henri Merke of a painting by George Webster

File:Die Seeschlacht bei Lissa.jpg *Source*: http://en.wikipedia.org/w/index.php?title=File:Die_Seeschlacht_bei_Lissa.jpg *License*: unknown *Contributors*: (1818-1879)

File:Novaral.jpg *Source*: http://en.wikipedia.org/w/index.php?title=File:Novaral.jpg *License*: unknown *Contributors*: Original uploader was Gdr at en.wikipedia

File:Boundary between Italy and Free Territory of Trieste.png *Source*: http://en.wikipedia.org/w/index.php?title=File:Boundary_between_Italy_and_Free_Territory_of_Trieste.png *License*: unknown *Contributors*: User:SreeBot

File:TrattatoDiOsimo.jpg *Source*: http://en.wikipedia.org/w/index.php?title=File:TrattatoDiOsimo.jpg *License*: unknown *Contributors*: User:Kiwicode

File:Dubrovnik view 01.JPG *Source*: http://en.wikipedia.org/w/index.php?title=File:Dubrovnik_view_01.JPG *License*: unknown *Contributors*: User:Luka Krstulović

File:Rimini Waterfront.jpg *Source*: http://en.wikipedia.org/w/index.php?title=File:Rimini_Waterfront.jpg *License*: unknown *Contributors*: User:RiminiCity

File:Portoroz-hotel Palace.JPG *Source*: http://en.wikipedia.org/w/index.php?title=File:Portoroz-hotel_Palace.JPG *License*: unknown *Contributors*: Original uploader was Ziga at sl.wikipedia

File:Triest Port1.JPG *Source*: http://en.wikipedia.org/w/index.php?title=File:Triest_Port1.JPG *License*: unknown *Contributors*: User:Welleschik

File:Luka brajdica 040408.jpg *Source*: http://en.wikipedia.org/w/index.php?title=File:Luka_brajdica_040408.jpg *License*: unknown *Contributors*: User:Roberta F.

File:Brosen durres panview.jpg *Source*: http://en.wikipedia.org/w/index.php?title=File:Brosen_durres_panview.jpg *License*: unknown *Contributors*: Albinfo, Brosen, Esc, Mac9

File:Flag of the People's Republic of China.svg *Source*: http://en.wikipedia.org/w/index.php?title=File:Flag_of_the_People's_Republic_of_China.svg *License*: unknown *Contributors*: User:Denelson83, User:SKopp, User:Shizhao, User:Zscout370

File:Flag of Thailand.svg *Source*: http://en.wikipedia.org/w/index.php?title=File:Flag_of_Thailand.svg *License*: unknown *Contributors*: User:Zscout370

File:Flag of the United States.svg *Source*: http://en.wikipedia.org/w/index.php?title=File:Flag_of_the_United_States.svg *License*: unknown *Contributors*: Anomie

File:Flag of India.svg *Source*: http://en.wikipedia.org/w/index.php?title=File:Flag_of_India.svg *License*: unknown *Contributors*: Anomie, Mifter

File:Flag of Saudi Arabia.svg *Source*: http://en.wikipedia.org/w/index.php?title=File:Flag_of_Saudi_Arabia.svg *License*: unknown *Contributors*: Unknown

GNU Free Documentation License Version 1.2, November 2002

Copyright (C) 2000,2001,2002 Free Software Foundation, Inc. 59 Temple Place, Suite 330, Boston, MA 02111-1307 USA Everyone is permitted to copy and distribute verbatim copies of this license document, but changing it is not allowed.

0. PREAMBLE

The purpose of this License is to make a manual, textbook, or other functional and useful document "free" in the sense of freedom: to assure everyone the effective freedom to copy and redistribute it, with or without modifying it, either commercially or noncommercially. Secondarily, this License preserves for the author and publisher a way to get credit for their work, while not being considered responsible for modifications made by others. This License is a kind of "copyleft", which means that derivative works of the document must themselves be free in the same sense. It complements the GNU General Public License, which is a copyleft license designed for free software. We have designed this License in order to use it for manuals for free software, because free software needs free documentation: a free program should come with manuals providing the same freedoms that the software does. But this License is not limited to software manuals; it can be used for any textual work, regardless of subject matter or whether it is published as a printed book. We recommend this License principally for works whose purpose is instruction or reference.

1. APPLICABILITY AND DEFINITIONS

This License applies to any manual or other work, in any medium, that contains a notice placed by the copyright holder saying it can be distributed under the terms of this License. Such a notice grants a world-wide, royalty-free license, unlimited in duration, to use that work under the conditions stated herein. The "Document", below, refers to any such manual or work. Any member of the public is a licensee, and is addressed as "you". You accept the license if you copy, modify or distribute the work in a way requiring permission under copyright law. A "Modified Version" of the Document means any work containing the Document or a portion of it, either copied verbatim, or with modifications and/or translated into another language. A "Secondary Section" is a named appendix or a front-matter section of the Document that deals exclusively with the relationship of the publishers or authors of the Document to the Document's overall subject (or to related matters) and contains nothing that could fall directly within that overall subject. (Thus, if the Document is in part a textbook of mathematics, a Secondary Section may not explain any mathematics.) The relationship could be a matter of historical connection with the subject or with related matters, or of legal, commercial, philosophical, ethical or political position regarding them. The "Invariant Sections" are certain Secondary Sections whose titles are designated, as being those of Invariant Sections, in the notice that says that the Document is released under this License. If a section does not fit the above definition of Secondary then it is not allowed to be designated as Invariant. The Document may contain zero Invariant Sections. If the Document does not identify any Invariant Sections then there are none. The "Cover Texts" are certain short passages of text that are listed, as Front-Cover Texts or Back-Cover Texts, in the notice that says that the Document is released under this License. A Front-Cover Text may be at most 5 words, and a Back-Cover Text may be at most 25 words. A "Transparent" copy of the Document means a machine-readable copy, represented in a format whose specification is available to the general public, that is suitable for revising the document straightforwardly with generic text editors or (for images composed of pixels) generic paint programs or (for drawings) some widely available drawing editor, and that is suitable for input to text formatters or for automatic translation to a variety of formats suitable for input to text formatters. A copy made in an otherwise Transparent file format whose markup, or absence of markup, has been arranged to thwart or discourage subsequent modification by readers is not Transparent. An image format is not Transparent if used for any substantial amount of text. A copy that is not "Transparent" is called "Opaque". Examples of suitable formats for Transparent copies include plain ASCII without markup, Texinfo input format, LaTeX input format, SGML or XML using a publicly available DTD, and standard-conforming simple HTML, PostScript or PDF designed for human modification. Examples of transparent image formats include PNG, XCF and JPG. Opaque formats include proprietary formats that can be read and edited only by proprietary word processors, SGML or XML for which the DTD and/or processing tools are not generally available, and the machine-generated HTML, PostScript or PDF produced by some word processors for output purposes only. The "Title Page" means, for a printed book, the title page itself, plus such following pages as are needed to hold, legibly, the material this License requires to appear in the title page. For works in formats which do not have any title page as such, "Title Page" means the text near the most prominent appearance of the work's title, preceding the beginning of the body of the text. A section "Entitled XYZ" means a named subunit of the Document whose title either is precisely XYZ or contains XYZ in parentheses following text that translates XYZ in another language. (Here XYZ stands for a specific section name mentioned below, such as "Acknowledgements", "Dedications", "Endorsements", or "History".) To "Preserve the Title" of such a section when you modify the Document means that it remains a section "Entitled XYZ" according to this definition. The Document may include Warranty Disclaimers next to the notice which states that this License applies to the Document. These Warranty Disclaimers are considered to be included by reference in this License, but only as regards disclaiming warranties: any other implication that these Warranty Disclaimers may have is void and has no effect on the meaning of this License.

2. VERBATIM COPYING

You may copy and distribute the Document in any medium, either commercially or noncommercially, provided that this License, the copyright notices, and the license notice saying this License applies to the Document are reproduced in all copies, and that you add no other conditions whatsoever to those of this License. You may not use technical measures to obstruct or control the reading or further copying of the copies you make or distribute. However, you may accept compensation in exchange for copies. If you distribute a large enough number of copies you must also follow the conditions in section 3. You may also lend copies, under the same conditions stated above, and you may publicly display copies.

3. COPYING IN QUANTITY

If you publish printed copies (or copies in media that commonly have printed covers) of the Document, numbering more than 100, and the Document's license notice requires Cover Texts, you must enclose the copies in covers that carry, clearly and legibly, all these Cover Texts: Front-Cover Texts on the front cover, and Back-Cover Texts on the back cover. Both covers must also clearly and legibly identify you as the publisher of these copies. The front cover must present the full title with all words of the title equally prominent and visible. You may add other material on the covers in addition. Copying with changes limited to the covers, as long as they preserve the title of the Document and satisfy these conditions, can be treated as verbatim copying in other respects. If the required texts for either cover are too voluminous to fit legibly, you should put the first ones listed (as many as fit reasonably) on the actual cover, and continue the rest onto adjacent pages. If you publish or distribute Opaque copies of the Document numbering more than 100, you must either include a machine-readable Transparent copy along with each Opaque copy, or state in or with each Opaque copy a computer-network location from which the general network-using public has access to download using public-standard network protocols a complete Transparent copy of the Document, free of added material. If you use the latter option, you must take reasonably prudent steps, when you begin distribution of Opaque copies in quantity, to ensure that this Transparent copy will remain thus accessible at the stated location until at least one year after the last time you distribute an Opaque copy (directly or through your agents or retailers) of that edition to the public. It is requested, but not required, that you contact the authors of the Document well before redistributing any large number of copies, to give them a chance to provide you with an updated version of the Document.

4. MODIFICATIONS

You may copy and distribute a Modified Version of the Document under the conditions of sections 2 and 3 above, provided that you release the Modified Version under precisely this License, with the Modified Version filling the role of the Document, thus licensing distribution and modification of the Modified Version to whoever possesses a copy of it. In addition, you must do these things in the Modified Version: A. Use in the Title Page (and on the covers, if any) a title distinct from that of the Document, and from those of previous versions (which should, if there were any, be listed in the History section of the Document). You may use the same title as a previous version if the original publisher of that version gives permission. B. List on the Title Page, as authors, one or more persons or entities responsible for authorship of the modifications in the Modified Version, together with at least five of the principal authors of the Document (all of its principal authors, if it has fewer than five), unless they release you from this requirement. C. State on the Title page the name of the publisher of the Modified Version, as the publisher. D. Preserve all the copyright notices of the Document. E. Add an appropriate copyright notice for your modifications adjacent to the other copyright notices. F. Include, immediately after the copyright notices, a license notice giving the public permission to use the Modified Version under the terms of this License, in the form shown in the Addendum below. G. Preserve in that license notice the full lists of Invariant Sections and required Cover Texts given in the Document's license notice. H. Include an unaltered copy of this License. I. Preserve the section Entitled "History", Preserve its Title, and add to it an item stating at least the title, year, new authors, and publisher of the Modified Version as given on the Title Page. If there is no section Entitled "History" in the Document, create one stating the title, year, authors, and publisher of the Document as given on its Title Page, then add an item describing the Modified Version as stated in the previous sentence. J. Preserve the network location, if any, given in the Document for public access to a Transparent copy of the Document, and likewise the network locations given in the Document for previous versions it was based on. These may be placed in the "History" section. You may omit a network location for a work that was published at least four years before the Document itself, or if the original publisher of the version it refers to gives permission. K. For any section Entitled "Acknowledgements" or "Dedications", Preserve the Title of the section, and preserve in the section all the substance and tone of each of the contributor acknowledgements and/or dedications given therein. L. Preserve all the Invariant Sections of the Document, unaltered in their text and in their titles. Section numbers or the equivalent are not considered part of the section titles. M. Delete any section Entitled "Endorsements". Such a section may not be included in the Modified Version. N. Do not retitle any existing section to be Entitled "Endorsements" or to conflict in title with any Invariant Section. O. Preserve any Warranty Disclaimers. If the Modified Version includes new front-matter sections or appendices that qualify as Secondary Sections and contain no material copied from the Document, you may at your option designate some or all of these sections as invariant. To do this, add their titles to the list of Invariant Sections in the Modified Version's license notice. These titles must be distinct from any other section titles. You may add a section Entitled "Endorsements", provided it contains nothing but endorsements of your Modified Version by various parties--for example, statements of peer review or that the text has been approved by an organization as the authoritative definition of a standard. You may add a passage of up to five words as a Front-Cover Text, and a passage of up to 25 words as a Back-Cover Text, to the end of the list of Cover Texts in the Modified Version. Only one passage of Front-Cover Text and one of Back-Cover Text may be added by (or through arrangements made by) any one entity. If the Document already includes a cover text for the same cover, previously added by you or by arrangement made by the same entity you are acting on behalf of, you may not add another; but you may replace the old one, on explicit permission from the previous publisher that added the old one. The author(s) and publisher(s) of the Document do not by this License give permission to use their names for publicity for or to assert or imply endorsement of any Modified Version.

5. COMBINING DOCUMENTS

You may combine the Document with other documents released under this License, under the terms defined in section 4 above for modified versions, provided that you include in the combination all of the Invariant Sections of all of the original documents, unmodified, and list them all as Invariant Sections of your combined work in its license notice, and that you preserve all their Warranty Disclaimers. The combined work need only contain one copy of this License, and multiple identical Invariant Sections may be replaced with a single copy. If there are multiple Invariant Sections with the same name but different contents, make the title of each such section unique by adding at the end of it, in parentheses, the name of the original author or publisher of that section if known, or else a unique number. Make the same adjustment to the section titles in the list of Invariant Sections in the license notice of the combined work. In the combination, you must combine any sections Entitled "History" in the various original documents, forming one section Entitled "History"; likewise combine any sections Entitled "Acknowledgements", and any sections Entitled "Dedications". You must delete all sections Entitled "Endorsements".

6. COLLECTIONS OF DOCUMENTS

You may make a collection consisting of the Document and other documents released under this License, and replace the individual copies of this License in the various documents with a single copy that is included in the collection, provided that you follow the rules of this License for verbatim copying of each of the documents in all other respects. You may extract a single document from such a collection, and distribute it individually under this License, provided you insert a copy of this License into the extracted document, and follow this License in all other respects regarding verbatim copying of that document.

7. AGGREGATION WITH INDEPENDENT WORKS

A compilation of the Document or its derivatives with other separate and independent documents or works, in or on a volume of a storage or distribution medium, is called an "aggregate" if the copyright resulting from the compilation is not used to limit the legal rights of the compilation's users beyond what the individual works permit. When the Document is included in an aggregate, this License does not apply to the other works in the aggregate which are not themselves derivative works of the Document. If the Cover Text requirement of section 3 is applicable to these copies of the Document, then if the Document is less than one half of the entire aggregate, the Document's Cover Texts may be placed on covers that bracket the Document within the aggregate, or the electronic equivalent of covers if the Document is in electronic form. Otherwise they must appear on printed covers that bracket the whole aggregate.

8. TRANSLATION

Translation is considered a kind of modification, so you may distribute translations of the Document under the terms of section 4. Replacing Invariant Sections with translations requires special permission from their copyright holders, but you may include translations of some or all Invariant Sections in addition to the original versions of these Invariant Sections. You may include a translation of this License, and all the license notices in the Document, and any Warranty Disclaimers, provided that you also include the original English version of this License and the original versions of those notices and disclaimers. In case of a disagreement between the translation and the original version of this License or a notice or disclaimer, the original version will prevail. If a section in the Document is Entitled "Acknowledgements", "Dedications", or "History", the requirement (section 4) to Preserve its Title (section 1) will typically require changing the actual title.

9. TERMINATION

You may not copy, modify, sublicense, or distribute the Document except as expressly provided for under this License. Any other attempt to copy, modify, sublicense or distribute the Document is void, and will automatically terminate your rights under this License. However, parties who have received copies, or rights, from you under this License will not have their licenses terminated so long as such parties remain in full compliance.

10. FUTURE REVISIONS OF THIS LICENSE

The Free Software Foundation may publish new, revised versions of the GNU Free Documentation License from time to time. Such new versions will be similar in spirit to the present version, but may differ in detail to address new problems or concerns. See http://www.gnu.org/copyleft/. Each version of the License is given a distinguishing version number. If the Document specifies that a particular numbered version of this License "or any later version" applies to it, you have the option of following the terms and conditions either of that specified version or of any later version that has been published (not as a draft) by the Free Software Foundation. If the Document does not specify a version number of this License, you may choose any version ever published (not as a draft) by the Free Software Foundation. ADDENDUM: How to use this License for your documents To use this License in a document you have written, include a copy of the License in the document and put the following copyright and license notices just after the title page: Copyright (c) YEAR YOUR NAME. Permission is granted to copy, distribute and/or modify this document under the terms of the GNU Free Documentation License, Version 1.2 or any later version published by the Free Software Foundation; with no Invariant Sections, no Front-Cover Texts, and no Back-Cover Texts. A copy of the license is included in the section entitled "GNU Free Documentation License". If you have Invariant Sections, Front-Cover Texts and Back-Cover Texts, replace the "with...Texts." line with this: with the Invariant Sections being LIST THEIR TITLES, with the Front-Cover Texts being LIST, and with the Back-Cover Texts being LIST. If you have Invariant Sections without Cover Texts, or some other combination of the three, merge those two alternatives to suit the situation. If your document contains nontrivial examples of program code, we recommend releasing these examples in parallel under your choice of free software license, such as the GNU General Public License, to permit their use in free software.

Printed by Books on Demand GmbH, Norderstedt / Germany